Lecture Notes in Mathematics 2163

More information about this series at http://www.springer.com/series/304

Viorel Barbu • Giuseppe Da Prato •
Michael Röckner

Stochastic Porous Media
Equations

 Springer

Viorel Barbu
Department of Mathematics
Al. I. Cuza University & Octav Mayer
  Institute of Mathematics of
  the Romanian Academy
Iasi, Romania

Michael Röckner
Department of Mathematics
University of Bielefeld
Bielefeld, Germany

Giuseppe Da Prato
Classe di Scienze
Scuola Normale Superiore di Pisa
Pisa, Italy

ISSN 0075-8434   ISSN 1617-9692   (electronic)
Lecture Notes in Mathematics
ISBN 978-3-319-41068-5   ISBN 978-3-319-41069-2   (eBook)
DOI 10.1007/978-3-319-41069-2

Library of Congress Control Number: 2016954369

Mathematics Subject Classification (2010): 60H15, 35K55, 76S99, 76M30, 76M35

This Springer imprint is published by Springer Nature
The registered company is Springer International Publishing AG Switzerland

Preface

This book is concerned with stochastic porous media equations with main emphasis on existence theory, asymptotic behaviour and ergodic properties of the associated transition semigroup. The general form of the porous media equation is

$$dX - \Delta\beta(X)dt = \sigma(X)dW, \qquad (1)$$

where $\beta : \mathbb{R} \to \mathbb{R}$ is a monotonically increasing function (possibly multivalued) and W is a cylindrical Wiener process.

Stochastic perturbations of the form $\sigma(X)\dot{W}$ in stochastic porous media equation were already considered by physicists but until recently no rigorous mathematical existence result was known. In specific models the noise arises from physical fluctuations of the media which in a first approximation can be taken of the form $(a + bX)\dot{W}$.

The porous media equation driven by a Gaussian noise, besides their relevance in the mathematical description of nonlinear diffusion dynamics perturbed by noise, has an intrinsic mathematical interest as a highly nonlinear partial differential equation, which is not well posed in standard spaces of regular functions. In fact the basic functional space for studying this equation is the distributional Sobolev space H^{-1} and this is due to the fact that the porous media operator $y \to -\Delta\beta(y)$ is m-accretive in the spaces H^{-1} and L^1 only. Since the Hilbertian structure of the space is essential for getting energetic estimates via Itô's formula, H^{-1} was chosen as an appropriate space for this equation.

Compared with the deterministic porous media equation which benefits from the theory of nonlinear semigroups of contractions in both the spaces L^1 and H^{-1}, the existence theory of the corresponding stochastic equations is not a direct consequence of general theory of the nonlinear Cauchy problem in Banach spaces. In fact, a nonlinear stochastic equation with additive noise (or with special linear noise) is formally equivalent with a nonlinear random differential equation with non-smooth time-dependent coefficients, which precludes the use of standard existence result for the deterministic Cauchy problem. However, the existence theory for stochastic infinite dimensional equations uses many techniques of nonlinear Cauchy problems associated with deterministic m-accretive nonlinear operators.

This book is organized into seven chapters. Chapter 1 is devoted to some standard topics from stochastic and nonlinear analysis mainly included without proof because they represent a necessary basic background for the subsequent topics.

Chapter 2 is devoted to existence theory for stochastic porous media equations with Lipschitz nonlinearity and may also be viewed as a background for the theory developed in Chap. 3, which is the core of the book. This chapter treats existence theory for equations with maximal monotone nonlinearities which have at most polynomial growths. The principal model described by this class of equations is the slow and fast diffusion processes. Besides existence, the extinction in finite time for fast diffusions and finite speed of propagation for slow diffusions are also studied.

Chapter 4 is devoted to the so-called variational approach to stochastic porous media equations. In a few words, the idea is to represent the equation as an infinite dimensional stochastic equation associated with a monotone and demi-continuous operator from a reflexive Banach space V to its dual V' and apply the standard existence theory developed in the early 1970s by E. Pardoux, N. Krylov and B. Rozovskii.

Chapter 5 is devoted to stochastic porous media equations with nonpolynomial growth to $\pm\infty$, for the diffusivity β, a situation which was excluded from the previous H^{-1} approach and which uses an L^1 treatment based on weak compactness arguments. The solution obtained in this way is weaker than in the previous case but applies to a larger class of functions β.

Chapter 6 is concerned with stochastic porous media equations in the whole $\mathbb{R}^d$.

Chapter 7 is devoted to existence and uniqueness of invariant measures for the transition semigroup associated with stochastic porous media equations.

These lecture notes have grown out of joint works and collaborations of authors in the last decade. They were written during their visits to Scuola Normale Superiore di Pisa and Bielefeld University.

Iasi, Romania   Viorel Barbu
Pisa, Italy   Giuseppe Da Prato
Bielefeld, Germany   Michael Röckner

Contents

1   **Introduction** 1
   1.1   Stochastic Porous Media Equations and Nonlinear Diffusions 1
      1.1.1   The Stochastic Stefan Two Phase Problem 5
   1.2   Preliminaries 6
      1.2.1   Functional Spaces and Notation 6
      1.2.2   The Gaussian Noise 8
      1.2.3   Stochastic Processes 9
      1.2.4   Monotone Operators 12

2   **Equations with Lipschitz Nonlinearities** 19
   2.1   Introduction and Setting of the Problem 19
      2.1.1   The Definition of Solutions 23
   2.2   The Uniqueness of Solutions 24
   2.3   The Approximating Problem 24
   2.4   Convergence of $\{X_\epsilon\}$ 28
      2.4.1   Estimates for $\|X_\epsilon(t)\|_{-1}^2$ 29
      2.4.2   Estimates for $\mathbb{E}\int_0^t \|F_\epsilon(X_\epsilon(s))\|_{-1}^2\,ds$ 31
      2.4.3   Additional Estimates in L^p 33
   2.5   The Solution to Problem (2.1) 37
   2.6   Positivity of Solutions 41
   2.7   Comments and Bibliographical Remarks 45

3   **Equations with Maximal Monotone Nonlinearities** 49
   3.1   Introduction and Setting of the Problem 49
   3.2   Uniqueness 51
   3.3   The Approximating Problem 53
      3.3.1   Estimating $\mathbb{E}\|X_\epsilon(t)\|_{-1}^2$ 54
      3.3.2   Estimating $\mathbb{E}|X_\epsilon(t)|_p^p$ 56
   3.4   Solution to Problem (3.1) 56
   3.5   Slow Diffusions 60
      3.5.1   The Uniqueness 61
   3.6   The Rescaling Approach to Porous Media Equations 63

3.7   Extinction in Finite Time for Fast Diffusions and Self
       Organized Criticality   65
3.8   The Asymptotic Extinction of Solutions to Self
       Organized Criticality   70
3.9   Localization of Solutions to Stochastic Slow Diffusion
       Equations: Finite Speed of Propagation   78
       3.9.1   Proof of Theorem 3.9.1   80
3.10   The Logarithmic Diffusion Equation   88
3.11   Comments and Bibliographical Remarks............................   93

4 Variational Approach to Stochastic Porous Media
   Equations   95
4.1   The General Existence Theory   95
4.2   An Application to Stochastic Porous Media Equations.............   98
4.3   Stochastic Porous Media Equations in Orlicz Spaces...............   99
4.4   Comments and Bibliographical Remarks............................   105

5 L^1-Based Approach to Existence Theory for Stochastic
   Porous Media Equations   107
5.1   Introduction and Setting of the Problem...........................   107
5.2   Proof of Theorem 5.1.4..   111
       5.2.1   A-Priori Estimates..   112
       5.2.2   Convergence for $\lambda \to 0$ 115
       5.2.3   Completion of Proof of Theorem 5.1.4......................   123
       5.2.4   Proof of Theorem 5.1.4   129
5.3   Comments and Bibliographical Remarks............................   131

6 The Stochastic Porous Media Equations in $\mathbb{R}^d$ 133
6.1   Introduction   133
6.2   Preliminaries   134
6.3   Equation (6.2) with a Lipschitzian β   137
6.4   Equation (6.2) for Maximal Monotone Functions β
       with Polynomial Growth   147
6.5   The Finite Time Extinction for Fast Diffusions   162
6.6   Comments and Bibliographical Remarks............................   165

7 Transition Semigroup   167
7.1   Introduction and Preliminaries   167
       7.1.1   The Infinitesimal Generator of P_t   169
7.2   Invariant Measures for the Slow Diffusions Semigroup P_t   170
7.3   Invariant Measure for the Stefan Problem   175
7.4   Invariant Measures for Fast Diffusions   178
       7.4.1   Existence..   178
       7.4.2   Uniqueness..   180
7.5   Invariant Measure for Self Organized Criticality Equation   186

7.6   The Full Support of Invariant Measures
        and Irreducibility of Transition Semigroups........................... 187
7.7   Comments and Bibliographical Remarks............................ 195

References... 197

Index... 201

Chapter 1
Introduction

This is an introductory chapter mainly devoted to the formulation of problems, models and some preliminaries on convex and infinite dimensional analysis, indispensable for understanding the sequel.

1.1 Stochastic Porous Media Equations and Nonlinear Diffusions

In this book we study nonlinear stochastic differential equations of the form

$$
\begin{cases}
dX(t) - \Delta\beta(X(t))dt = \sigma(X(t))dW(t), & \text{in } [0, +\infty) \times \mathcal{O}, \\[2mm]
\beta(X(t)) = 0, & \text{on } [0, +\infty) \times \partial\mathcal{O}, \\[2mm]
X(0) = x, & \text{in } \mathcal{O},
\end{cases}
\tag{1.1}
$$

where $\mathcal{O}$ is an open (bounded) domain of $\mathbb{R}^d$, $d \geq 1$, with boundary $\partial\mathcal{O}$, β is a (multivalued) maximal monotone function from $\mathbb{R}$ to itself, W is a Wiener process and $\sigma = \sigma(X)$ is a suitable function to be made precise later on.

The deterministic equation

$$
\begin{cases}
\dfrac{\partial}{\partial t}X(t) - \Delta\beta(X(t))dt = f(t, \xi), & \text{in } [0, +\infty) \times \mathcal{O}, \\[3mm]
\beta(X(t)) = 0, & \text{on } [0, +\infty) \times \partial\mathcal{O}, \\[2mm]
X(0) = x, & \text{in } \mathcal{O},
\end{cases}
\tag{1.2}
$$

© Springer International Publishing Switzerland 2016
V. Barbu et al., *Stochastic Porous Media Equations*, Lecture Notes
in Mathematics 2163, DOI 10.1007/978-3-319-41069-2_1

is referred to in the literature as the *porous media equation* because the first model described by (1.2) was the dynamics of the flow in a porous medium. (In this case $\beta(X) = X^m$, $m > 1$). The stochastic equation (1.1) can be seen as an extension of (1.2) when the forcing term is replaced by a noise term.

It should be said that equations of the form (1.1) describe a large class of nonlinear diffusion mathematical models we briefly describe below.

If we formally represent (1.1) as

$$dX(t) - \text{div } (\beta'(X(t))\nabla X(t))dt = \sigma(X(t))dW(t), \qquad (1.3)$$

then we recognize in (1.3) the classical diffusion equation with diffusion coefficient $j(r) = \int_0^r \beta(s)ds$. In all diffusion models $X(t)$ represents the *mass concentration* and so (1.3) can be viewed as the mathematical model for the dynamics of diffusion flows driven by a stochastic perturbation $\sigma(X(t))dW(t)$.

As mentioned earlier, the standard model of diffusion of a gas in porous media is that where $\beta(r) = |r|^{m-1}r$ with $m > 1$ which is also called the *slow diffusion* model. More generally, we can consider the case where β is a continuous monotone function satisfying

$$\rho|r|^{m+1} \le r\beta(r) \le b_1|r|^{q+1} + b_2r, \quad \forall\, r \in \mathbb{R}, \qquad (1.4)$$

for $m > 1$ and $q > m$, ρ, $b_1 > 0$.

The case $m = 2$ describes the flow of an ideal gas in porous media while $m \ge 2$ that of a diffusion of a compressible fluid through porous media. There are other situations such as thermal propagation in plasma ($m = 6$) or plasma radiation ($m = 4$) which are modelled by the same equation. Some models in population dynamics are represented by (1.3) for $\beta(r) = ar^2$.

The case when $0 < m < 1$ is that of *fast diffusion* and is relevant in the description of plasma physics, the kinetic theory of gas or fluid transportation in porous media (see [28, 29, 42, 89]). As a matter of fact, these equations are associated with superdiffusive processes in which the time growth of the mean square displacement $\langle X^2(t) \rangle$ has a nonlinear growth in time as $t \to \infty$.

The singular case $-1 < m < 0$ or $\beta(r) = \log r$ in the limiting case $m = 0$ in the equation

$$dX(t) - \text{div } (u^{m-1}\nabla u)dt = \sigma(X(t))dW(t) \qquad (1.5)$$

models the *superfast diffusion* arising in the description of dynamics of plasma in magnetic fields as well as in the central limit approximation to the Carleman model of the Boltzmann equation [28, 29, 88]. In 2-D the corresponding deterministic equation describes the evolution of a surface by the Ricci curvature flow (see e.g. [90].)

A general feature of the fast and superfast diffusion models is that they model diffusion processes with a fast speed of mass transportation and this is one reason for which, as we shall see later on, the process terminates within finite time with positive probability.

The *self-organized criticality* (SOC) equation is the special case of (1.1) where

$$\beta(r) = \rho \operatorname{sign} r + \nu(r), \quad \forall\, r \in \mathbb{R}, \tag{1.6}$$

where $\rho > 0$, ν is a maximal monotone graph in $\mathbb{R} \times \mathbb{R}$ and

$$\operatorname{sign} r = \begin{cases} \frac{r}{|r|} & \text{for } r \neq 0, \\[2mm] \{r \in \mathbb{R} : |r| = 1\} & \text{for } r = 0. \end{cases} \tag{1.7}$$

In this case Eq. (1.1), that is

$$dX(t) - \rho\, \Delta(\operatorname{sign} X(t))dt + \nu(X(t))dt \ni \sigma(X(t))dW(t), \tag{1.8}$$

is used as a mathematical model for the standard self-organized criticality process, called the *sand-pile model* or the *Bak–Tang–Wiesenfeld (BTW) model* [4, 5] and which can be formalized via the cellular automaton model briefly presented below.

Consider an $N \times N$ square matrix representing a spacial discrete region $\mathscr{O} = \{X_{i,j}\}$, $i,j = 1, \ldots, N$. To each site (i,j) is assigned at moment t a nonnegative (integer) variable $X_{ij}(t)$. The dynamics of the $\mathbb{R}_+^{N^2}$-valued variable $X(t) = \{X_{i,j}(t)\}$, $i,j = 1, \ldots, N$, is described by the equation

$$X_{i,j}(t+1) \to X_{i,j}(t) - Z_{kl}^{ij} \quad \text{for } (k,l) \in \Gamma_{ij}, \tag{1.9}$$

where $\Gamma_{ij} = \{(i+1,j), (i,j+1), (i-1,j), (i,j-1)\}$ is the set of all four nearest neighbors of (i,j) and

$$Z_{kl}^{ij} = \begin{cases} 4 & \text{if } i = k,\, j = l, \\ -1 & \text{if } (k,l) \in \Gamma_{ij}, \\ 0 & \text{if } (k,l) \notin \Gamma_{ij}. \end{cases}$$

The algebraic law (1.9) describes rigorously what happens with the "activated" site (i,j) (i.e. a site which has attained or is over the critical height X_c): it looses four grains of sands which move to nearest neighbors in the interval of time $(t, t+1)$. This is a small "avalanche" which leads to a new configuration of the sand-pile. This transition from $X(t)$ to $X(t+1)$ can be written as

$$X_{i,j}(t+1) - X_{i,j}(t) = -Z_{ij}H(X_{i,j}(t) - X_c), \quad i,j = 1, \ldots, N, \tag{1.10}$$

where H is the Heaviside function

$$H(r) = \begin{cases} 1 & \text{if } r > 0, \\ 0 & \text{if } r < 0 \end{cases}$$

and $Z_{ij} = Z_{ij}^{kl}$, $k, l \in \Gamma_{ij}$. We assume here that the boundary sites (i, j) are in the subcritical case, that is $X_{i,j} - (X_\epsilon)_{i,j} \leq 0$.

The exact meaning of (1.10) is that the transfer dynamics (1.9) works in the critical or supercritical region only i.e., in an activated site (i, j), where $X_{ij} > X_c$. Otherwise, we can consider that X_{ij} remains unchanged. We can represent it as

$$X(t+1) - X(t) = -ZH(X(t) - X_c), \tag{1.11}$$

where $Z = Z_{ij}$, $i, j = 1, \ldots, N$. It can be seen that Z_{ij} is the second order difference operator in the spacial domain $\mathcal{O}$, i.e.

$$Z_{ij}(Y_{ij}) = Y_{i+1,j} + Y_{i-1,j} + Y_{i,j+1} - 4Y_{i,j} + Y_{i,j-1} \quad \forall\, i, j = 1, \ldots, N \tag{1.12}$$

and so, Eq. (1.12) is the discrete version of the partial differential equation of parabolic type

$$\begin{cases} \dfrac{\partial}{\partial t} X(t) = \Delta H(X(t) - X_c) & \text{in } \mathcal{O}, \\[2mm] X(t) - X_c = 0 & \text{on } \partial\mathcal{O}, \\[2mm] X(0) = x & \text{in } \mathcal{O}. \end{cases}$$

where Δ is the Laplace operator.

Therefore, if we replace the finite region $\mathcal{O}$ by a continuous domain in 2-dimensional space (for instance $\mathcal{O} = (0, 1) \times (0, 1)$) and the site location (i, j) by a point ξ in $\mathcal{O}$ the above model reduces to a nonlinear diffusion equation on the spatial domain $\mathcal{O} \subset \mathbb{R}^2$.

In the literature there are several versions of (SOC) equations modelling sandpile processes. One of this is

$$\begin{cases} \dfrac{\partial}{\partial t}(X(t) = \Delta X H(X(t) - X_c)) & \text{in } \mathcal{O}, \\[2mm] X(t) - X_c = 0 & \text{on } \partial\mathcal{O}, \\[2mm] X(0) = x & \text{in } \mathcal{O}. \end{cases}$$

If one perturbs this spontaneous process by a stochastic process of the form $\sigma(X(t))W(t)$ which, roughly speaking, means that one adds grains of sand to Gaussian random locations one gets an equation of type (1.8).

It must be said that there are other SOC models described by superfast diffusion equations of the form (1.1), (1.2) where $\beta(r) = \rho r^m$, $-2 < m < 0$ (see [28, 29]). But the mathematical treatment of such a problem remains to be done.

It should be emphasized also that the self-organized Eq. (1.8) can be viewed itself as a fast diffusion equation if we take into account that at least formally it can be written as

$$dX(t) - \rho \, \text{div} \, (\delta(X(t))\nabla X(t))dt + \nu(X(t))dt \ni \sigma(X(t))dW(t), \tag{1.13}$$

where δ is the Dirac measure concentrated in the origin. This reveals the complexity and high singularity of Eq. (1.8).

1.1.1 The Stochastic Stefan Two Phase Problem

In the special case

$$\beta(r) = \begin{cases} ar & \text{for } r < 0, \\ 0 & \text{for } 0 \le r \le \rho, \\ b(r - \rho) & \text{for } r > \rho, \end{cases} \tag{1.14}$$

$a, b, \rho > 0$, $\sigma(r) = \beta(r)$, Eq. (1.1) that is

$$\begin{cases} dX(t) - \Delta\beta(X(t))dt = \beta(X(t))dW(t), & \text{in } [0, +\infty) \times \mathcal{O}, \\ \\ \beta(X(t)) = 0, & \text{on } [0, +\infty) \times \partial\mathcal{O}, \\ \\ X(0) = x, & \text{in } \mathcal{O}, \end{cases} \tag{1.15}$$

reduces to the two phase transition Stefan problem perturbed by a Gaussian noise. More precisely

$$\begin{cases} d\theta - a\Delta\theta dt = \theta dW(t) & \text{in } \{(t, \xi) : \theta(t, \xi) < 0\}, \\ \\ d\theta - b\Delta\theta dt = \theta dW(t) & \text{in } \{(t, \xi) : \theta(t, \xi) > 0\}, \\ \\ (a\nabla\theta^+ - b\nabla\theta^-) \cdot \nabla\ell = -\rho & \text{in } \{(t, \xi) : \theta(t, \xi) = 0\}, \end{cases} \tag{1.16}$$

where

$$\{(t, \xi) : \theta(t, \xi) = 0\} = \{(t, \xi) : t = \ell(\xi)\}.$$

Here $\theta = \theta(t, \xi)$ is the temperature and $X = \beta^{-1}(\theta)$ is the enthalpy of the system (see e.g. [55]). This mathematical model describes the situation where the melting or solidification process is driven by a stochastic heat flow $\theta\dot{W}(t))$ which is proportional to the temperature. We note that (1.16) is a stochastic partial differential equation with free (moving) boundary $\{(t, \xi) : \theta(t, \xi) = 0\}$ which involves a transmission condition on the free boundary. Other phase transition diffusion models (for instance the oxygen diffusion in an absorbing tissue) are described by similar equations.

The Stefan one phase problem of the form (1.16) was intensively studied in last years in 1-D (see for instance [69, 70] and the references given there). It should be mentioned however that such a problem is a stochastic variational inequality which can not be represented in the form (1.15) and so its treatment is beyond the scope of this work.

1.2 Preliminaries

1.2.1 Functional Spaces and Notation

Let $\mathcal{O}$ be a bounded open subset of $\mathbb{R}^d$. We assume that its boundary $\partial \mathcal{O}$ is sufficiently regular (at least of class C^2) in order to apply the paper [66].

The following spaces will be considered in what follows.

- $L^p(\mathcal{O}) = L^p$, $p \in [1, \infty]$, is the Banach space of all p-summable (equivalence classes of) functions on $\mathcal{O}$ with the usual norm $| \cdot |_p$. The inner product in the Hilbert space $L^2(\mathcal{O})$ will be denoted by $\langle \cdot, \cdot \rangle_2$.
- $H^k(\mathcal{O}) = H^k$, $k \in \mathbb{N}$, is the Sobolev space of all functions in L^2 whose distributional derivatives of order lesser than k belong to L^2. $H_0^1(\mathcal{O}) = H_0^1$ is the set of all functions of H^1 that vanish on the boundary of $\mathcal{O}$. The norm in H_0^1 is denoted by $\| \cdot \|_1$ and given by

$$\|u\|_1 := \left(\int_{\mathbb{R}^d} |\nabla u|_{\mathbb{R}^d}^2 \, d\xi \right)^{1/2}.$$

- $\mathscr{D}'(\mathcal{O})$ denotes the space of Schwartz distributions on $\mathcal{O}$.
- $H^{-1}(\mathcal{O}) = H^{-1}$ is the dual of $H_0^1(\mathcal{O})$. Its norm will be denoted by $\| \cdot \|_{-1}$ and the inner product by $\langle \cdot, \cdot \rangle_{-1}$.
- Given a Banach space Z and $1 \leq p \leq \infty$, $W^{1,p}([0, T]; Z)$ denotes the space of all absolutely continuous functions $u : [0, T] \to Z$ which are a.e. differentiable on $[0, T]$ and $\frac{du}{dt} \in L^p(0, T; Z)$.

It is well known that the linear operator $A : H_0^1(\mathcal{O}) \to H^{-1}(\mathcal{O})$

$$Ax = -\Delta x, \quad \forall x \in H_0^1(\mathcal{O}), \tag{1.17}$$

is continuous and positive definite, while its restriction to $L^2(\mathcal{O})$ with domain $D(A) = H_0^1(\mathcal{O}) \cap H^2(\mathcal{O})$ is self-adjoint. Moreover, for all $x, y \in H^{-1}(\mathcal{O})$ we have

$$\langle x, y \rangle_{-1} = \langle A^{-1/2} x, A^{-1/2} y \rangle_2.$$

If $y \in L^2$ then,

$$\langle x, y \rangle_{-1} = \langle A^{-1}x, y \rangle_2.$$

The operator $A : D(A) \subset L^2(\mathcal{O}) \to L^2(\mathcal{O})$ where $D(A) = H^2(\mathcal{O}) \cap H_0^1(\mathcal{O})$, is symmetric and possesses a complete orthonormal basis of eigenfunctions $\{e_h\}_{h \in \mathbb{N}}$. We denote by $(\alpha_h)_{h \in \mathbb{N}}$ the corresponding set of eigenvalues,

$$\Delta e_h = -\alpha_h e_h, \quad \forall h \in \mathbb{N}. \tag{1.18}$$

We note that, setting $f_h = (\alpha_h)^{1/2} e_h$, $h \in \mathbb{N}$, then $\{f_h\}_{h \in \mathbb{N}}$ is a complete orthonormal basis of H^{-1}.

Example 1.2.1 Let $\mathcal{O} = (0,1)^d$ then we have

$$e_h(\xi) = (2/\pi)^{-\frac{d}{2}} \sin(h_1 \pi \xi_1) \cdots \sin(h_d \pi \xi_d), \quad h_1, \ldots, h_d \in \mathbb{N},$$

$$f_h(\xi) = (2/\pi)^{-\frac{d}{2}} \pi |h| \sin(h_1 \pi \xi_1) \cdots \sin(h_d \pi \xi_d)), \quad h_1, \ldots, h_d \in \mathbb{N},$$

and

$$\alpha_h = \pi^2 |h|^2,$$

where $|h|^2 = h_1^2 + \cdots + h_d^2$. $\qquad\square$

It is useful to specify the asymptotic behavior of e_h, f_h and α_h as $h \to \infty$, when $(0,1)^d$ is replaced by an arbitrary open set $\mathcal{O} \subset \mathbb{R}^d$ with smooth boundary. First we note that there exists $a_\mathcal{O} > 0$ such that

$$\alpha_0^{-1} h^{\frac{2}{d}} \le a_h \le a_\mathcal{O} h^{\frac{2}{d}}, \quad \forall h \in \mathbb{N}, \tag{1.19}$$

where the enumeration is chosen in increasing order counting multiplicity. (see e.g. [73, 91], [67, Vol. 3, Corollary 17.8.5] and the references therein.)

Moreover, there exists $b_\mathcal{O} > 0$ such that

$$|e_h(\xi)| \le b_\mathcal{O} \alpha_h^{\frac{d-1}{2}}, \quad |f_h(\xi)| \le b_\mathcal{O} \alpha_h^{\frac{d}{2}}, \quad \forall h \in \mathbb{N}, \xi \in \mathcal{O}. \tag{1.20}$$

(see [66].) This estimate is optimal for general sets $\mathcal{O}$. Obviously in the particular case where $\mathcal{O} = (0,1)^d$ we have a better result

$$|e_h(\xi)| \le (2/\pi)^{-\frac{d}{2}}, \quad |f_h(\xi)| \le (2/\pi)^{-\frac{d}{2}} \pi |h|, \quad \forall h \in \mathbb{N}^d, \xi \in \mathcal{O}. \tag{1.21}$$

1.2.2   The Gaussian Noise

Let H be a separable Hilbert space, $\{g_h\}$ an orthonormal basis of H and let $(W_h)_{h\in\mathbb{N}}$ be a sequence of independent real $\mathscr{F}_t$-Brownian motions on a filtered probability space $(\Omega, \mathscr{F}, \mathbb{P}, (\mathscr{F}_t)_{t\geq 0})$ for some normal filtration $(\mathscr{F}_t)_{t\geq 0}$. This means that W_h is $\mathscr{F}_t$-adapted and $W_h(t + a) - W_h(a)$ is independent of $\mathscr{F}_t$ for any $h \in \mathbb{N}$ and any $t, a \geq 0$. Then we define the *cylindrical Wiener process* in H as the formal series

$$W(t) := \sum_{k\in\mathbb{N}} W_h(t)g_h.$$

Though for any $t > 0$ this series is not convergent in $L^2(\Omega, \mathscr{F}, \mathbb{P})$, it is easily seen that the series

$$BW(t) := \sum_{k\in\mathbb{N}} W_h(t)Bg_h,$$

is convergent in $L^2(\Omega, \mathscr{F}, \mathbb{P}; C([0, T]; H))$ if and only if $B \in \mathscr{L}_2(H)$. Here $\mathscr{L}_2(H)$ is the space of all Hilbert–Schmidt operators in H endowed with the norm

$$\|T\|_{\mathscr{L}_2(H)} = (\mathrm{Tr}\,[TT^*])^{1/2}, \quad T \in \mathscr{L}_2(H)$$

and $\mathrm{Tr}\,[TT^*]$ represents the trace of TT^*,

$$\mathrm{Tr}\,[TT^*] = \mathrm{Tr}\,[T^*T] = \sum_{k\in\mathbb{N}} \|Tg_k\|_{-1}^2,$$

for one (and consequently every) orthonormal basis $\{g_k\}$ of H.

Given a progressively measurable process $F : [0, T] \times \Omega \to \mathscr{L}_2(H)$ such that $\mathbb{E}\int_0^T \|F(s)\|_{\mathscr{L}_2(H)}^2\,ds < \infty$, the Itô integral

$$\int_0^T F(s)dW(s) = \sum_{k=1}^{\infty} \int_0^T F(s)g_k dW_k(s)$$

is a well defined random variable in $L^2(\Omega, \mathscr{F}, \mathbb{P}; H)$ (see e.g. [51, 82]) and we have

$$\mathbb{E}\left\| \int_0^T F(s)dW(s) \right\|_{-1}^2 = \int_0^T \mathbb{E}\|F(s)\|_{\mathscr{L}_2(H)}^2\,ds \qquad (1.22)$$

1.2.3 Stochastic Processes

Let again H be a separable Hilbert space and $p, q \in [1, +\infty]$. We consider the following spaces of stochastic processes $X : [0, T] \times \Omega \to H$.

- $L_W^q(0, T; L^p(\Omega, H))$ is the space of all progressively measurable processes such that

$$\int_0^T [\mathbb{E}(|X(t)|^p)]^q dt < \infty.$$

- $C_W([0, T]; L^p(\Omega, H))$ is the space of all H-valued progressively measurable processes which are p mean square continuous on $[0, T]$, that is such that

$$\sup_{t \in [0,T]} \mathbb{E}(|X(t)|^p) < \infty.$$

- $L_W^2(\Omega; C([0, T]; H))$ is the space of all H-valued progressively measurable processes which are continuous on $[0, T]$ and such that

$$\mathbb{E} \sup_{t \in [0,T]} |X(t)|^2 < \infty.$$

It is well known that there is a natural imbedding of $L_W^2(\Omega; C([0, T]; H))$ into $C_W([0, T]; L^p(\Omega, H))$.

Let $F \in L_W^2([0, T]; L^2(\Omega, \mathscr{L}_2(H))$ and set

$$X(t) := \int_0^t F(s) dW(s), \quad t \in [0, T].$$

(Everywhere in the following the stochastic integral is considered in the sense of Itô.) Then X belongs to $C_W([0, T]; L^2(\Omega, H))$ and possesses a version which belongs to $L_W^2(\Omega; C([0, T]; H))$. Moreover, X is a *martingale* and the following result holds:

Proposition 1.2.2 (Burkholder–Davis–Gundy) *For arbitrary $p > 0$ there exists a constant $c_p > 0$ such that*

$$\mathbb{E} \left(\sup_{t \leq T} \left| \int_0^t F(s) dW(s) \right|^p \right) \leq c_p \mathbb{E} \left(\left[\int_0^T \|F(s)\|_{\mathscr{L}_2(H)}^2 \, ds \right]^{p/2} \right). \tag{1.23}$$

(see e.g. [51, Theorem 4.36]).

By Doob's maximal inequality we have

$$\mathbb{E} \sup_{t \in [0,T]} \left| \int_0^t F(s) dW(s) \right|^2 \leq 4 \int_0^T \mathbb{E} \|F(s)\|_{\mathscr{L}_2(H)}^2 \, ds. \tag{1.24}$$

An *Itô's process* with values in H is a stochastic process $X(t)$, $t \in [0,T]$, of the form

$$X(t) = x + \int_0^t b(s)ds + \int_0^t \sigma(s)dW(s), \qquad (1.25)$$

where $x \in H$, $b \in L^1_W([0,T]; L^1(\Omega, H))$ and $\sigma \in L^2_W([0,T]; L^2(\Omega, \mathscr{L}_2(H)))$. Then if $\varphi \in C^2(H)$ (i.e. twice continuously Fréchet differentiable on H) the Itô formula holds (see e.g. [38, Theorem 2.4], [51]),

$$\varphi(X(t)) = \varphi(x) + \int_0^t \langle b(s), D\varphi(X(s)) \rangle ds$$

$$+ \frac{1}{2} \int_0^t \mathrm{Tr}\, [\sigma^*(s)D^2\varphi(X(s))\sigma(s)]ds \qquad (1.26)$$

$$+ \int_0^t \langle D\varphi(X(s)), \sigma(s)dW(s) \rangle,$$

where $D\varphi$ and $D^2\varphi$ represent the first and second derivatives of φ, respectively. Identity (1.26) holds in $L^2_W(\Omega; C([0,T]; H))$, $\mathbb{P}$-a.s.

We note for further use that

$$\mathrm{Tr}\, [\sigma^*(s)D^2\varphi(X(s))\sigma(s)] = \sum_{k=1}^{\infty} \langle D^2\varphi(X(s))\sigma(s)g_k, \sigma(s)g_k \rangle, \qquad (1.27)$$

where $\{g_k\}$ is an orthonormal basis of H.

1.2.3.1 Itô's Formula for the L^p Norm

Here we present a result on Itô's formula for the L^p norm due to Krylov [71]. Let $(\Omega, \mathscr{F}, \mathscr{F}_t, \mathbb{P})$ be a filtered probability space as before and let denote by $\mathscr{P}$ the corresponding predictable σ-algebra in $\Omega \times [0, +\infty)$.

We consider processes

$$u : \Omega \times [0, +\infty) \to L^p(\mathbb{R}^d),$$

which satisfy the equation

$$du(t) = f(t)dt + g_j(t)dW_j(t), \qquad (1.28)$$

where f, g_j, $1 \leq j < \infty$ are L^p-valued processes. (Here we have used the summation convention over repeated indices.)

If τ is a stopping time we set for a separable Banach space E

$$\mathbb{L}^p(\tau, E) = L^p([0, \tau], \widetilde{\mathscr{P}}, L^p(\mathbb{R}^d, E)),$$

where $\widetilde{\mathscr{P}}$ is the completion of $\mathscr{P}$ with respect to $\mathbb{P}(d\omega) \times dt$. We have [71, Lemma 5.1].

Proposition 1.2.3 *Let* $2 \leq p < \infty$, $f \in \mathbb{L}^p(\tau, \mathbb{R})$, $g = \{g_k\} \in \mathbb{L}^p(\tau, \ell^2)$, *and let* u *be a progressively measurable map on* $\Omega \times [0, \infty)$ *with values in the space of distributions on* $\mathbb{R}^d$ *such that for any* $\varphi \in C_0^\infty(\mathbb{R}^d)$ *with probability one for all* $t \in (0, +\infty)$ *we have*

$$\langle u(t \wedge \tau), \varphi \rangle_2 = \langle u_0, \varphi \rangle_2 + \int_0^t \mathbb{1}_{s \leq \tau} \langle f(s), \varphi \rangle_2 ds$$

$$+ \sum_{k=1}^\infty \int_0^t \langle g_k(s), \varphi \rangle_2 \mathbb{1}_{s \leq \tau} dW_k(s),$$

(1.29)

where $u_0 \in L^p(\Omega, \mathscr{F}_0, L^p)$.[1] *Then there is* $\Omega' \subset \mathscr{F}_0$ *of full probability such that*

(i) $u(t \wedge \tau) \mathbb{1}_{\Omega'}$ *is an* L^p-*valued adapted continuous process on* $[0, +\infty)$.
(ii) *For all* $t \in [0, +\infty)$ *and* $\omega \in \Omega'$, *we have*

$$|u(t \wedge \tau)|_p^p = |u_0|_p^p + \int_0^{t \wedge \tau} \left(p \int_{\mathbb{R}^d} |u(s)|^{p-2} u(s) f(s) dx \right.$$

$$+ \frac{1}{2} p(p-1) \int_{\mathbb{R}^d} |u(s)|^{p-2} \sum_{k=1}^\infty |g_k(s)|^2 dx \bigg) ds$$

(1.30)

$$+ p \int_0^{t \wedge \tau} \int_{\mathbb{R}^d} |u(s)|^{p-2} u(s) \sum_{k=1}^\infty g_k(s) dx \, dW_k(s).$$

We note also that Proposition 1.2.3 remains true on domains $\mathcal{O} \subset \mathbb{R}^d$ by replacing u with $u \mathbb{1}_{\mathcal{O}}$.

The proof of Proposition 1.2.3 is given in the above quoted paper by Krylov [71]. Here we confine only to point out the main steps of the proof.

First it turns out that we may replace u in (1.29) by a function measurable with respect to

$$\mathscr{F} \times \mathscr{B}([0, s]) \times \mathscr{B}(\mathbb{R}^d)$$

---

[1] Here we use $\langle \cdot, \cdot \rangle_2$ also to denote the duality between $C_0^\infty(\mathbb{R}^d)$ and the space of distributions.

such that for each ξ, $u(t,\xi)$ is $\mathscr{F}_t$-adapted, $u(t,\xi,\omega)$ is continuous in $t \in (0,\infty)$ for each $(\omega,\xi) \in \Omega \times \mathbb{R}^d$ and $u(t,\cdot,\omega)$ as a function of (t,ω) is L^p-valued, $\mathscr{F}_t$-adapted and continuous in t for each ω.

Next consider a mollifier $\rho_\epsilon(\xi) = \epsilon^{-d}\rho(\frac{\xi}{\epsilon})$ and set

$$f_\epsilon(t) = (f * \rho_\epsilon)(t), \quad u_\epsilon(t) = (u * \rho_\epsilon)(t), \quad t \geq 0.$$

We obtain the equation

$$u_\epsilon(t,\xi) = (u_0)_\epsilon(\xi) + \int_0^t f_\epsilon(s,\xi)ds + \int_0^t (g_j)_\epsilon(s,\xi)dW_j(s).$$

Since u_ϵ is regular we may apply Itô's formula and get $\mathbb{P}$-a.s.

$$|u_\epsilon(t)|_p^p = |(u_0)_\epsilon|_p^p + p\int_0^t |u_\epsilon(s)|^{p-2}u_\epsilon(s)(g_j)_\epsilon(s)dW_j(s)$$

$$+ \int_0^t p|u_\epsilon(s)|^{p-2}u_\epsilon(s)f_\epsilon(s)ds$$

$$+ \tfrac{1}{2}p(p-1)\int_0^t |u_\epsilon(s)|^{p-2}u(s)\sum_{k=1}^\infty |(g_k)_\epsilon(s)|ds.$$

Then formula (1.30) follows after some a priori estimates involving the stochastic Fubini theorem and letting $\epsilon \to 0$.

We shall apply the following version of a martingale convergence result (see e.g.[75, p. 139]).

Lemma 1.2.4 *Let Z be a nonnegative semimartingale with $\mathbb{E}(Z(t)) < \infty, \forall t \geq 0$ and let I be a nondecreasing continuous process such that*

$$Z(t) + I(t) = Z(0) + I_1(t) + M(t), \quad \forall t \geq 0, \tag{1.31}$$

where M is a local martingale. Then if $\lim_{t\to\infty} I_1(t) < \infty$, $\mathbb{P}$-a.s., we have

$$\lim_{t\to\infty} Z(t) + I(\infty) < \infty, \quad \mathbb{P}\text{-a.s.} \tag{1.32}$$

1.2.4 Monotone Operators

Let H be a real Hilbert space with the scalar product $\langle\cdot,\cdot\rangle$ and norm $|\cdot|$. A multivalued mapping $G: D(G) \subset H \to 2^H$ is called *monotone* if

$$\langle u - v, x - y \rangle \geq 0, \quad \forall x \in G(u), \, y \in G(v).$$

A monotone mapping G is called *maximal monotone* if $1 + \alpha G$ is surjective for all $\alpha > 0$ (Equivalently for some $\alpha > 0$.) If G is maximal monotone we set

$$J_\alpha(x) = (1 + \alpha G)^{-1}(x), \quad \alpha > 0, \ x \in H.$$

(Here 1 is the identity operator in H.)

Lemma 1.2.5 J_α *is Lipschitzian and*

$$|J_\alpha(x) - J_\alpha(y)| \leq |x - y|, \quad \forall \, x, y \in H. \tag{1.33}$$

Proof Set $x_\alpha = J_\alpha(x)$, $y_\alpha = J_\alpha(y)$, so that

$$x = x_\alpha + \alpha G(x_\alpha), \quad y = y_\alpha + \alpha G(y_\alpha).$$

Then

$$x_\alpha - y_\alpha + \alpha(G(x_\alpha) - G(y_\alpha)) = x - y.$$

Multiplying both sides by $x_\alpha - y_\alpha$ and taking into account the accretivity of G, yields

$$|x_\alpha - y_\alpha|^2 \leq \langle x_\alpha - y_\alpha, x - y \rangle$$

and the conclusion follows easily.   □

We define the *Yosida approximations* $G_\alpha : H \to H$ of G setting for any $\alpha > 0$.

$$G_\alpha = \frac{1}{\alpha}(1 - J_\alpha). \tag{1.34}$$

Since J_α is 1-Lipschitz, it follows that G_α is $\frac{2}{\alpha}$-Lipschitz on H.

Proposition 1.2.6 *Let $\alpha > 0$. Then we have*

(i) $G_\alpha(x) \in G(J_\alpha(x))$, $\quad \forall \, x \in H$.
(ii) $|G_\alpha(x)| \leq |G^0(x)|$, $\quad \forall \, x \in D(G)$, *where $G^0(x)$ is the minimal section of $G(x)$.*

Proof

(i) Let $\alpha > 0$ and $x \in H$. Then

$$G_\alpha(x) = \frac{1}{\alpha}[(1 - \alpha G)(J_\alpha(x)) - J_\alpha(x)] = G(J_\alpha(x)).$$

(ii) We have

$$G_\alpha(x) = \frac{1}{\alpha}[J_\alpha(x)(x - \alpha y) - J_\alpha(x)], \quad \forall \, y \in G(x).$$

Since J_α is 1-Lipschitz, it follows that

$$|G_\alpha(x)| \leq \frac{1}{\alpha} |x - (x - \alpha y)| = |y|, \quad \forall\, y \in G(x)$$

and (ii) follows.   □

If $\varphi : H \to \overline{\mathbb{R}} := (-\infty, +\infty]$ is a convex and lower semicontinuous function we denote its subdifferential by $\partial\varphi$, that is

$$\partial\varphi(x) = \{y \in H : \varphi(x) \leq \varphi(u) + \langle y, x - u\rangle, \ \forall\, u \in H\}. \tag{1.35}$$

Then $\partial\varphi : H \to H$ is maximal monotone and its Yosida approximation $(\partial\varphi)_\alpha$ is monotone, Lipschitz and it is given by

$$(\partial\varphi)_\alpha(x) = \nabla\varphi_\alpha(x), \quad \forall\, x \in H, \tag{1.36}$$

where the convex function $\varphi_\alpha : H \to \mathbb{R}$, defined by

$$\varphi_\alpha(x) = \inf\left\{\varphi(y) + \frac{|x - y|^2}{2\alpha} : y \in H\right\}$$

$$\tag{1.37}$$

$$= \varphi((1 + \alpha\partial\varphi)^{-1}x) + \frac{\alpha}{2}|x - (1 + \alpha\partial\varphi)^{-1}x|^2,$$

is the *Moreau* regularization of φ.

More generally, if X is a Banach space with dual X', the operator $G : X \to 2^{X'}$ is said to be *maximal monotone* if it is monotone, that is

$$(u - v, x - y) \geq 0, \quad \forall\, u \in G(x), \ v \in G(y),$$

and $\alpha F + G : X \to X'$ is surjective for all $\alpha > 0$. Here $F : X \to X'$ is the duality mapping of X and $(\cdot, \cdot)$ is the duality pairing of X, X'. A maximal monotone operator $G : X \to X'$ is strongly–weakly closed, that is if $y_n \in G(x_n)$ and $x_n \to x$ strongly in X and $y_n \to y$ weakly in X' then $y \in G(x)$. Also G is weakly–strongly closed. We note that if $G : X \to X'$ is monotone and demicontinuous (that is strongly–weakly continuous) it is maximal monotone (Minty–Browder theorem).

An important example of a maximal monotone operator is the *subdifferential* $\partial\varphi : X \to 2^{X'}$ of a convex lower semicontinuous function, $\varphi : X \to \overline{\mathbb{R}} = (-\infty, +\infty]$, that is

$$\partial\varphi(x) = \{y \in X' : \varphi(x) \leq \varphi(u) + (y, x - u), \quad \forall\, u \in X\}.$$

If $\beta : \mathbb{R} \to 2^{\mathbb{R}}$ is a maximal monotone mapping (graph), that is

$$(u - v)(x - y) \geq 0, \quad \forall\, u \in \beta(x), \ v \in \beta(y),$$

and $(1 + \beta)(\mathbb{R}) = \mathbb{R}$, then there is a unique convex lower semicontinuous function $j : \mathbb{R} \to \mathbb{R}$ such that $\partial J = \beta$. This function j is called the *potential* of β. (The uniqueness of j is up to additive constants.)

Let j^* denote the conjugate of j (the Legendre transform of j), that is

$$j^*(p) = \sup\{py - j(y) : y \in \mathbb{R}\}$$

We recall that $(\partial j)^* = \partial j^{-1}$ (see e.g. [14]),

$$j(y) + j^*(p) = py \quad \text{if and only if } p \in \partial j(y) \tag{1.38}$$

and

$$j(y) + j^*(p) \geq py \quad \text{for all } j, p \in \mathbb{R}. \tag{1.39}$$

If β_α, $\alpha > 0$, is the Yosida approximation of β we set

$$j_\alpha(u) = \int_0^u \beta_\alpha(r)dr, \quad u \in \mathbb{R}$$

and note that j_α is just the Moreau approximation of j, that is

$$j_\alpha(u) = \min\left\{ j(v) + \frac{1}{2\alpha}|u - v|^2, \quad v \in \mathbb{R} \right\}. \tag{1.40}$$

We have

$$j_\alpha(u) = (1 + \alpha\beta)^{-1}(u) + \frac{1}{2\alpha}|u - (1 + \alpha\beta)^{-1}(u)|^2. \tag{1.41}$$

Below we present a few examples of maximal monotone operators (see [6, 35])

Example 1.2.7 Let $g : \mathbb{R} \to (-\infty, +\infty]$ be a lower semicontinuous convex function and let $\varphi : L^p(\widetilde{\mathscr{O}}) \to \overline{\mathbb{R}}$ be defined by

$$\varphi(y) = \begin{cases} \int_{\mathscr{O}} g(y(\xi))\mu(d\xi), & \text{if } g(y) \in L^1(\widetilde{\mathscr{O}}) \\ +\infty & \text{otherwise,} \end{cases} \tag{1.42}$$

where $\widetilde{\mathscr{O}}$ is a measure space endowed with a σ-finite measure μ and $1 \leq p < \infty$. Then φ is convex, lower semicontinuous on $L^p(\widetilde{\mathscr{O}})$ and

$$\partial\varphi(y) = \{z \in L^q(\widetilde{\mathscr{O}}) : z(\xi) \in \partial g(y(\xi)), \ \mu\text{-a.e. } \xi \in \widetilde{\mathscr{O}}\}, \tag{1.43}$$

where $p^{-1} + q^{-1} = 1$, see e.g. [14]. Moreover, the conjugate $\varphi^* \colon L^2(\mathscr{O}) \to \overline{\mathbb{R}}$ of φ,

$$\varphi^*(z) = \sup_y \left\{ \int_{\mathscr{O}} zy d\mu - \varphi(y) \right\}$$

is given by

$$\varphi^*(z) = \int_{\mathscr{O}} g^*(z(\xi))\mu(d\xi),$$

where g^* is the conjugate of g.

   The following simple lemma is a very useful tool to pass to the limit in nonlinear equations.

Proposition 1.2.8 *Let β be a maximal monotone graph in $\mathbb{R} \times \mathbb{R}$ and let $\{u_n\} \subset L^p(\widetilde{\mathscr{O}})$, $\{v_n\} \subset L^q(\widetilde{\mathscr{O}})$, $\frac{1}{p} + \frac{1}{q} = 1$, be such that*

$$\lim_{n\to\infty} u_n = u \text{ weakly in } L^p(\widetilde{\mathscr{O}}), \quad \lim_{n\to\infty} v_n = v \text{ weakly in } L^q(\widetilde{\mathscr{O}}),$$

where $v_n \in \beta(u_n)$ μ-a.e. on $\widetilde{\mathscr{O}}$. Assume that

$$\limsup_{n\to\infty} \int_{\mathscr{O}} v_n u_n d\mu \leq \int_{\mathscr{O}} vu d\mu. \tag{1.44}$$

Then $v(\xi) \in \beta(u(\xi))$, μ-a.e. $\xi \in \widetilde{\mathscr{O}}$.

Proof We have for all $y \in L^p(\mathscr{O})$, $z \in L^q(\mathscr{O})$ such that $z \in \beta(y)$, μ-a.e. in $\widetilde{\mathscr{O}}$ that

$$\int_{\widetilde{\mathscr{O}}} (v_n - z)(u_n - y) d\mu \geq 0, \quad \forall n \in \mathbb{N}$$

by the monotonicity of β. Letting $n \to \infty$ it follows by (1.44) that

$$\int_{\widetilde{\mathscr{O}}} (v - z)(u - y) d\mu \geq 0.$$

   Since β is a maximal monotone graph, we may choose $(y, z) \in L^p(\widetilde{\mathscr{O}}) \times L^q(\widetilde{\mathscr{O}})$ such that

$$y + z = u + v, \quad z \in \beta(y), \ \mu\text{-a.s. in } \widetilde{\mathscr{O}}.$$

(We may assume here that $p \geq q$ and so $y = (1 + \beta)^{-1}(u + v) \in L^p(\widetilde{\mathscr{O}})$.) This yields $v(\xi) \in \beta(u(\xi))$ for μ-a.e. $\xi \in \widetilde{\mathscr{O}}$ as claimed. $\square$

   More generally, Proposition 1.2.8 extends to maximal monotone operators $G \colon X \to X'$. Namely one has, see [7, p. 38]

Proposition 1.2.9 *If $\{u_n\} \subset X$, $\{v_n\} \subset X'$ are such that $v_n \in G(u_n)$, $u_n \to u$ weakly in X, $v_n \to v$ weakly in X' and $\limsup_{n\to\infty} \langle v_n, u_n \rangle \leq \langle v, u \rangle$, then $v \in G(u)$.*

Example 1.2.10 Let $\mathscr{O}$ be an open and bounded subset of $\mathbb{R}^d$ and let $g : \mathbb{R} \to (-\infty, +\infty]$ be a lower semicontinuous convex function and let $\varphi : L^2(\mathscr{O}) \to \overline{\mathbb{R}}$ be defined by

$$
\varphi(y) = \begin{cases} \int_{\mathscr{O}} \left(\frac{1}{2}|\nabla y(\xi)|^2 + g(y(\xi))\right) d\xi, & \text{if } g(y) \in L^1(\mathscr{O}), \ y \in H^1_0(\mathscr{O}), \\ +\infty & \text{otherwise.} \end{cases} \tag{1.45}
$$

Assume that $\partial\mathscr{O}$ is of class C^2. Then φ is convex, lower semicontinuous on $L^2(\mathscr{O})$ and

$$
\begin{cases} D(\partial\varphi) = \{y \in H^1_0\mathscr{O}) \cap H^2(\mathscr{O}) : \ \partial g(y) \cap L^2(\mathscr{O}) \neq \varnothing\}, \\ \partial\varphi(y) = \{z \in L^2(\mathscr{O}) : z(\xi) \in -\Delta y(\xi) + \partial g(y(\xi)), \ \text{a.e. } \xi \in \mathscr{O}\}. \end{cases} \tag{1.46}
$$

Example 1.2.11 Let $\mathscr{O}$ be an open and bounded subset of $\mathbb{R}^d$ and let $g : \mathbb{R} \to (-\infty, +\infty]$ be a lower semicontinuous convex function such that

$$
\lim_{r\to\infty} \frac{g(r)}{r} = +\infty. \tag{1.47}
$$

Let $\beta = \partial g$. Then the operator $F : H^{-1}(\mathscr{O}) \to H^{-1}(\mathscr{O})$ defined by

$$
\begin{cases} F(y) = -\Delta\beta(y), \quad \forall y \in D(F), \\ D(F) := \{y \in H^{-1}(\mathscr{O}) \cap L^1(\mathscr{O}) : \ \exists \xi \in \beta(y), \ \text{a.e. in } \mathscr{O}, \ \xi \in H^1_0(\mathscr{O})\}, \end{cases} \tag{1.48}
$$

is maximal monotone in $H^{-1}(\mathscr{O})$ and $F = \partial\varphi$ where $\varphi : H^{-1}(\mathscr{O}) \to (-\infty, +\infty]$ is given by

$$
\varphi(y) = \begin{cases} \int_{\mathscr{O}} g(y(\xi))d\xi & \text{if } g(y) \in L^1(\mathscr{O}), \\ +\infty & \text{otherwise.} \end{cases} \tag{1.49}
$$

We note that condition (1.47) is equivalent with $\beta(\mathbb{R}) = \mathbb{R}$. See [6].

Let $(\mathscr{O}_0, \mathscr{F}, d\xi)$ be a measure space with finite measure $d\xi$ and let $\mathscr{U} \subset L^1(\mathscr{O}_0)$ be a family of integrable functions. We say that $\mathscr{U}$ is *equi-integrable* if for any $\epsilon > 0$

there is $\delta > 0$ such that for any measurable set $G \subset \mathscr{O}_0$ with $|G| < \delta$ it follows that

$$\int_G |u|\,d\xi \le \epsilon \quad \forall\, u \in \mathscr{U}.$$

(Here $|G| = \int_G d\xi$.)

We have the following classical result

Theorem 1.2.12 (Dunford–Pettis) *If $\mathscr{U}$ is bounded in $L^1(\mathscr{O}_0)$ and equi-integrable it is weakly sequentially compact in $L^1(\mathscr{O}_0)$.*

Chapter 2
Equations with Lipschitz Nonlinearities

We start here by studying the porous media equation problem (1.1) when β : $\mathbb{R} \to \mathbb{R}$ is monotonically increasing and Lipschitz continuous. The main reason is that general maximal monotone graphs β can be approximated by their Yosida approximations β_ϵ which are Lipschitz continuous and monotonically increasing. So, several estimates proved in this chapter will be exploited later for studying problems with more general β.

We also note that there are significant physical problems with β Lipschitz continuous as in the case of the Stefan two phase problem presented in Sect. 1.1.1.

2.1 Introduction and Setting of the Problem

Everywhere in the following we shall simply write L^p, H_0^1 and H^{-1} instead of $L^p(\mathcal{O})$, $H_0^1(\mathcal{O})$ and $H^{-1}(\mathcal{O})$ respectively. The corresponding norms will be denoted $|\cdot|_2$, $\|\cdot\|_1$, $\|\cdot\|_{-1}$ and the scalar products by $\langle\cdot,\cdot\rangle_2$, $\langle\cdot,\cdot\rangle_1$ and $\langle\cdot,\cdot\rangle_{-1}$. If $u \in H_0^1$ and $v \in H^{-1}$ we shall denote by $\langle\cdot,\cdot\rangle$ the duality from H_0^1 and H^{-1}. (It coincides with $\langle\cdot,\cdot\rangle_2$ on $H_0^1 \times L^2$.) By $_p\langle\cdot,\cdot\rangle_q$ we shall denote the duality pairing between L^p and L^q, $p^{-1} + q^{-1} = 1$.

As in Sect. 1.2.1 $\{e_k\}$ represents an orthonormal basis of eigenfunctions of $-\Delta$ in L^2 and $\{\alpha_k\}$ the corresponding sequence of eigenvalues,

$$-\Delta e_k = \alpha_k e_k \text{ in } \mathcal{O}, \quad e_k = 0 \text{ on } \partial\mathcal{O}.$$

We set $f_k = \alpha_k^{1/2} e_k$ so that $\{f_k\}$ is an orthonormal basis in H^{-1}. The domain $\mathcal{O} \subset \mathbb{R}^d$ is assumed to be bounded and with a boundary $\partial\mathcal{O}$ of class C^2.

© Springer International Publishing Switzerland 2016
V. Barbu et al., *Stochastic Porous Media Equations*, Lecture Notes
in Mathematics 2163, DOI 10.1007/978-3-319-41069-2_2

We are here concerned with the following stochastic differential equation in H^{-1}

$$\begin{cases} dX(t) = \Delta\beta(X(t))dt + \sigma(X(t))dW(t), \\ \\ X(0) = x \in H^{-1}, \end{cases} \tag{2.1}$$

were Δ is the Laplace operator with Dirichlet homogeneous boundary conditions. We write Eq. (2.1) equivalently as

$$\begin{cases} dX(t) = -F(X(t))dt + \sigma(X(t))dW(t), \\ \\ X(0) = x \in H^{-1}, \end{cases} \tag{2.2}$$

where F is defined by

$$\begin{cases} F(x) := -\Delta\beta(x), \quad \forall x \in D(F), \\ \\ D(F) = \{x \in L^1 \cap H^{-1} : \beta(x) \in H_0^1\} \end{cases} \tag{2.3}$$

If β is Lipschitz continuous and $\beta(0) = 0$ then F maps $D(F)$ into H^{-1}. In general F is monotone in H^{-1}. In fact if $x, \bar{x} \in D(F)$ we have

$$\langle F(x) - F(\bar{x}), x - \bar{x}\rangle_{-1} = \langle -\Delta\beta(x) + \Delta\beta(\bar{x}), x - \bar{x}\rangle_{-1}$$

$$= \langle \beta(x) - \beta(\bar{x}), x - \bar{x}\rangle_2 \geq 0.$$

If in addition $\beta(\mathbb{R}) = \mathbb{R}$ then F is maximal monotone because the condition (1.47) from Example 1.2.11 is in that case fulfilled.

Let us make precise our assumptions on β and σ.

Hypothesis 1

(i) $\beta : \mathbb{R} \to \mathbb{R}$ is monotonically increasing and $\beta(0) = 0$. Moreover, there exists $K > 0$ such that

$$|\beta(r) - \beta(s)| \leq K|r - s|, \quad \forall r, s \in \mathbb{R}. \tag{2.4}$$

(ii) σ is Lipschitzian from H^{-1} into $\mathscr{L}_2(H^{-1})$
 As a consequence there exists $K_1 > 0$ such that

$$\|\sigma(x) - \sigma(y)\|_{\mathscr{L}_2(H^{-1}, H^{-1})} \leq K_1 \|x - y\|_{-1}, \quad \forall x, y \in H^{-1}, \tag{2.5}$$

and we have

$$\|\sigma(x)\|_{\mathscr{L}_2(H^{-1}, H^{-1})} \leq \sigma_1 + K_1 \|x\|_{-1}, \quad \forall x \in H^{-1}, \tag{2.6}$$

where $\sigma_1 = \|\sigma(0)\|_{\mathscr{L}_2(H^{-1}, H^{-1})}$

(iii) σ is Lipschitzian from L^2 into $\mathcal{L}_2(H^{-1}, L^2)$.[1]

As a consequence there exists $K_2 > 0$ such that

$$\|\sigma(x) - \sigma(y)\|^2_{\mathcal{L}_2(H^{-1}, L^2)} \leq K_2 |x - y|^2_2, \quad \forall\, x, y \in L^2, \tag{2.7}$$

and we have

$$\|\sigma(x)\|_{\mathcal{L}_2(H^{-1}, L^2)} \leq \sigma_2 + K_2 |x|_2, \quad \forall\, x \in L^2, \tag{2.8}$$

where $\sigma_2 := |\sigma(0)|_{\mathcal{L}_2(H^{-1}, L^2)}$.

(iv) W is a cylindrical Wiener process in H^{-1}.

Several remarks about Hypothesis 1 are in order. Concerning (i), we note, for further use, the following obvious but useful consequence,

$$(\beta(r) - \beta(s))(r - s) \geq \frac{1}{K} (\beta(r) - \beta(s))^2, \quad \forall\, r, s \in \mathbb{R}. \tag{2.9}$$

As regards the cylindrical Wiener process in H^{-1} (see Sect. 1.2.2) we shall thus have

$$W(t) := \sum_{k \in \mathbb{N}} W_k(t) f_k, \tag{2.10}$$

where $(W_k)_{k \in \mathbb{N}}$ is a sequence of independent real Brownian motions on a filtered probability space $(\Omega, \mathscr{F}, \mathbb{P}, (\mathscr{F}_t)_{t \geq 0})$ and $\{f_k\}$ is the orthonormal basis of H^{-1} introduced after (1.18).

We note that (ii) and (iii) allow us to use Itô's formula in H^{-1} and L^2 respectively.

Finally, we shall see that, as far as existence and uniqueness of solutions on H^{-1} are concerned, condition (iii) can be dropped.

Let us finish this section with a few examples.

Example 2.1.1 (Additive Noise) We assume here that $\sigma = \sqrt{Q}$ with $Q \in \mathcal{L}_1(H^{-1})^2$. In this case condition (ii) in Hypothesis 1 is obviously fulfilled.

Assume in particular that $Q = A^{-\gamma}$. Therefore (ii) is fulfilled provided

$$\sum_{k=1}^{\infty} \alpha_k^{-\gamma} < \infty$$

---

[1] Here $\mathcal{L}_2(H^{-1}, L^2)$ denotes the space of all Hilbert–Schmidt operators from H^{-1} to L^2.

[2] $\mathcal{L}_1(H^{-1})$ denotes the space of all symmetric, nonnegative definite, trace-class operators in H^{-1}.

and so by (1.19) provided $\gamma > \frac{d}{2}$. (iii) is fulfilled provided

$$\gamma > \frac{d+2}{2} \tag{2.11}$$

□

Example 2.1.2 (Linear Noise) Here we take W as in (2.10) and $\sigma(x)$ depending linearly on x,

$$\sigma(x)h = KxA^{-\gamma/2}h = K\sum_{k=1}^{\infty}(\alpha_k)^{-\gamma/2}(xf_k)h_k, \tag{2.12}$$

for all x, $h \in H^{-1}$, where $h_k = \langle h, f_k \rangle_{-1}$ and γ is sufficiently large and $K \in [0, \infty)$.
To choose γ, let us recall the following analytic inequalities

$$\|xf_k\|_{-1}^2 \le C_1\alpha_k^d \|x\|_{-1}^2, \quad \forall\, k \in \mathbb{N}. \tag{2.13}$$

and

$$|xe_k|_2^2 \le C_2\alpha_k^{d-1} |x|_2^2, \quad \forall\, k \in \mathbb{N}, \tag{2.14}$$

proved in Appendix A below.
Now by (2.13), taking into account (1.19), we have

$$\|\sigma(x)\|_{\mathscr{L}_2(H^{-1})}^2 = K^2 \sum_{k=1}^{\infty} \|xA^{-\gamma/2}f_k\|_{-1}^2 \le C_1 K^2 \sum_{k=1}^{\infty} \alpha_k^{d-\gamma} \|x\|_{-1}^2 < +\infty,$$

provided $\gamma > \frac{3}{2}d$, and (ii) holds in this case.
Concerning (iii) we have

$$|\sigma(x)|_{\mathscr{L}_2(H^{-1},L^2)}^2 = K^2\|xA^{-\gamma/2}\|_{\mathscr{L}_2(H^{-1},L^2)}^2$$

$$= K^2 \sum_{k=1}^{\infty} \alpha_k^{1-\gamma} |xe_k|_2^2 \le C_2 \sum_{k=1}^{\infty} \alpha_k^{d-\gamma} |x|_2^2,$$

where we used (1.20). So (recalling (1.19)) we see that (iii) is fulfilled provided $\gamma > \frac{3}{2}d$.
In conclusion, for the mapping σ defined by (2.12), Hypothesis 1 is fulfilled when

$$\gamma > \frac{3}{2}d. \tag{2.15}$$

Example 2.1.3 (Stefan Problem) We consider β defined by (1.14), that is

$$\beta(r) = \begin{cases} ar & \text{for } r < 0, \\ 0 & \text{for } 0 \le r \le \rho, \\ b(r - \rho) & \text{for } r > \rho, \end{cases} \tag{2.16}$$

where $a, b, \rho > 0$. We notice that β fulfills Hypothesis 1-(i) with $L = \min\{a, b\}$.

As seen in Sect. 1.1.1, in this case Eq. (2.1) models the *Stefan two phase* heat transfer (melting solidification) in presence of a stochastic Gaussian perturbation. Here β is the inverse of the *enthalpy* function associated with the phase transition and X is related with the temperature θ by the transformation $\theta = \beta(X)$.

2.1.1 The Definition of Solutions

From now on until the end of Chap. 2 we assume that Hypothesis 1 is in force.

Since F is monotone and σ is Lipschitz in H^{-1}, it is natural to look for solutions of problem (2.1) in H^{-1}. Then, as we shall see, the well posedness of problem (2.1) will follow from general properties of monotone operators provided an a priori estimate for $\|F(X(t))\|_{-1}$ is available.

Definition 2.1.4 For any $x \in H^{-1}$ a *strong solution* to (2.1) in $[0, T]$ is a stochastic process X which satisfies

$$X \in L_W^2(\Omega; C([0, T]; H^{-1})) \cap L^2(0, T; L^2(\Omega; L^2)),$$

$\beta(X) \in L^2(0, T; L^2(\Omega; H_0^1))$ and

$$X(t) = x + \int_0^t \Delta\beta(X(s))ds + \int_0^t \sigma(X(s))dW(s), \quad \forall\, t \in [0, T], \; \mathbb{P}\text{-a.s.} \tag{2.17}$$

It follows from (2.17) that a strong solution of (2.1) is an Itô process in H^{-1}.

For $x \in H^{-1}$ a *generalized solution* to (2.1) in $[0, T]$ is a process X which belongs to $L_W^2(\Omega; C([0, T]; H^{-1}))$ and there exists a sequence $\{x_n\} \subset L^2$ convergent to x in H^{-1} and such that

$$\lim_{n \to \infty} X_n = X \quad \text{in } L_W^2(\Omega; C([0, T]; H^{-1})),$$

where X_n is the strong solution to (2.1) with x_n replacing x.

Clearly for $x \in L^2$ the concepts of strong and generalized solution agree.

2.2   The Uniqueness of Solutions

The uniqueness of strong or generalized solutions is a standard consequence of the monotonicity of F and the Lipschitzianity of σ in H^{-1}.

Proposition 2.2.1 *Equation* (2.1) *has at most one strong solution.*

Proof Let first X and Y be two strong solutions to (2.1). Then by Itô's formula we have

$$
\begin{aligned}
d\|X(t) - Y(t)\|_{-1}^2 &= 2\langle X(t) - Y(t), \Delta(\beta(X(t))) - \Delta(\beta(Y(t)))\rangle_{-1}\, dt \\
&= 2\langle X(t) - Y(t), (\sigma(X(t)) - \sigma(Y(t)))dW(t)\rangle_{-1} \qquad (2.18) \\
&\quad + \|\sigma(X(t)) - \sigma(Y(t))\|_{\mathscr{L}_2(H^{-1})}^2\, dt, \qquad \mathbb{P}\text{-a.s.}
\end{aligned}
$$

Therefore

$$
\begin{aligned}
&d\|X(t) - Y(t)\|_{-1}^2 + 2\langle X(t) - Y(t), \beta(X(t)) - \beta(Y(t))\rangle_2\, dt \\
&+ 2\langle X(t) - Y(t), (\sigma(X(t)) - \sigma(Y(t)))dW(t)\rangle_{-1} \qquad (2.19) \\
&+ \|\sigma(X(t)) - \sigma(Y(t))\|_{\mathscr{L}_2(H^{-1})}^2\, dt, \qquad \mathbb{P}\text{-a.s.}
\end{aligned}
$$

Taking into account assumption (2.5), the monotonicity of β and taking expectation, yields

$$
\mathbb{E}\|X(t) - Y(t)\|_{-1}^2 \le \mathbb{E}\int_0^t \|(\sigma(X(s)) - \sigma(Y(s)))\|_{\mathscr{L}_2(H^{-1})}^2 ds
$$

$$
\le K_1 \mathbb{E}\int_0^t \|X(s) - Y(s)\|_{-1}^2 ds.
$$

Now the conclusion follows from Gronwall's lemma.   □

2.3   The Approximating Problem

The basic method used here to prove existence for Eq. (1.1) is to approximate it by an equation of the form

$$
\begin{cases}
dX_\epsilon + F_\epsilon(X_\epsilon)dt = \sigma(X_\epsilon)dW(t), \\[2mm]
X_\epsilon(0) = x,
\end{cases}
\qquad (2.20)
$$

where $\epsilon \in (0, 1)$, F_ϵ (to be defined later on), is Lipschitz continuous in H^{-1}, so that problem (2.20) is well posed by the classical theory of SPDEs, see [51].

When $\beta(\mathbb{R}) = \mathbb{R}$, then (as noticed earlier) F is maximal monotone and in this case a natural choice would be to take F_ϵ to be the Yosida approximation of F. To avoid this restrictive conditions we shall proceed as follows. First we consider the operator

$$G_\epsilon(x) := -\Delta(\beta + \epsilon I)(x), \quad \forall x \in H_0^1, \tag{2.21}$$

which is maximal monotone in H^{-1} because $(\beta + \epsilon I)(\mathbb{R}) = \mathbb{R}$, (see Example 1.2.11), where here and below I denotes the identity map on the respective space. Then we define $J_\epsilon : H^{-1} \to H_0^1$ by

$$J_\epsilon := (I + \epsilon G_\epsilon)^{-1}, \quad \epsilon > 0, \tag{2.22}$$

and

$$F_\epsilon(x) := \frac{1}{\epsilon}(x - J_\epsilon(x)) = -\Delta(\beta + \epsilon I)(J_\epsilon(x)), \quad x \in H^{-1}. \tag{2.23}$$

So, problem (2.20) is equivalent to

$$\begin{cases} dX_\epsilon = \Delta(\beta + \epsilon I)(J_\epsilon(X_\epsilon))dt + \sigma(X_\epsilon)dW(t), \\ X_\epsilon(0) = x. \end{cases} \tag{2.24}$$

We are going to show that $F_\epsilon = -\Delta(\beta + \epsilon I)(J_\epsilon(X_\epsilon))$ is Lipschitz continuous both in H^{-1} and in L^2 so that problem (2.20) has a unique strong solution X_ϵ belonging to $L_W^2(\Omega; C([0, T]; H^{-1}))$ for all $x \in H^{-1}$ and a unique strong solution X_ϵ belonging to $L_W^2(\Omega; C([0, T]; L^2))$ for all $x \in L^2$ by standard existence results (see e.g. [51].)

Let us prove some preliminaries.

Lemma 2.3.1 *Let $\epsilon > 0$.*

(i) We have

$$\|J_\epsilon(x)\|_{-1} \leq \|x\|_{-1}, \quad \forall x \in H^{-1}, \tag{2.25}$$

$$\|J_\epsilon(x) - J_\epsilon(x_1)\|_{-1} \leq \|x - x_1\|_{-1}, \quad \forall x, x_1 \in H^{-1}, \tag{2.26}$$

$$|J_\epsilon(x)|_2 \leq |x|_2, \quad \forall x \in L^2 \tag{2.27}$$

and

$$|J_\epsilon(x) - J_\epsilon(x_1)|_2 \leq \frac{c}{\epsilon}|x - x_1|_2, \quad \forall x, x_1 \in L^2, \tag{2.28}$$

for a suitable $c > 0$. Consequently, J_ϵ and hence F_ϵ are both Lipschitz continuous both in H^{-1} and in L^2. Furthermore, $J_\epsilon(x) \in H_0^1$ for all $x \in H^{-1}$.
(ii) Let $p > 2$. Then for all $x \in L^p$

$$|J_\epsilon(x)|_p \le |x|_p \tag{2.29}$$

and

$$\int_{\mathcal{O}} |x|^{p-2} x F_\epsilon(x) \, d\xi \ge 0. \tag{2.30}$$

Proof (i): Since $\beta(0) = 0$, (2.25) follows from (2.26) which we prove now. Let $x, x_1 \in H^{-1}$ and set $y = J_\epsilon(x)$, $y_1 = J_\epsilon(x_1)$. Then we have

$$(y - y_1) - \epsilon(\Delta(\beta + \epsilon I)(y) + (\Delta(\beta + \epsilon I)(y_1)) = x - x_1.$$

Multiplying scalarly in H^{-1} both sides of the identity above by $y - y_1$ and taking into account the dissipativity of F, yields

$$\|y - y_1\|_{-1}^2 \le \|y - y_1\|_{-1} \|x - x_1\|_{-1},$$

from which (2.26) follows.

Let us show (2.28) (we note that F is not monotone in L^2). Given $x, \bar{x} \in L^2$ set

$$J_\epsilon(x) = y, \quad J_\epsilon(\bar{x}) = \bar{y},$$

so that

$$y - \bar{y} - \epsilon \Delta(\beta + \epsilon I)(y) + \epsilon \Delta(\beta + \epsilon I)(\bar{y}) = x - \bar{x}. \tag{2.31}$$

Note that $(\beta + \epsilon I)^{-1}$ is Lipschitz continuous, because

$$|(\beta + \epsilon I)^{-1}(r) - (\beta + \epsilon I)^{-1}(s)| \le \epsilon^{-1} |r - s|, \quad \forall \, r, s \in \mathbb{R}. \tag{2.32}$$

As a consequence, since $(\beta + \epsilon I)(y) \in H_0^1$, we have $y \in H_0^1$. Multiplying both sides of (2.31) by $(\beta + \epsilon I)(y) - (\beta + \epsilon I)(\bar{y})$ and integrating over $\mathcal{O}$, yields

$$\int_{\mathcal{O}} (y - \bar{y})((\beta + \epsilon I)(y) - (\beta + \epsilon I)(\bar{y})) d\xi$$

$$-\epsilon \, \langle (\Delta(\beta + \epsilon I)(y) - \Delta(\beta + \epsilon I)(\bar{y})), ((\beta + \epsilon I)(y)$$

$$-(\beta + \epsilon I)(\bar{y})) \rangle$$

$$= \int_{\mathcal{O}} (x - \bar{x})((\beta + \epsilon I)(y) - (\beta + \epsilon I)(\bar{y})) d\xi.$$

Taking into account that β is monotone increasing it follows that

$$\epsilon |y - \bar{y}|_2^2 + \epsilon \|(\beta + \epsilon I)(y) - (\beta + \epsilon I)(\bar{y})\|_1^2 \le |x - \bar{x}|_2 \, |(\beta + \epsilon I)(y) - (\beta + \epsilon I)(\bar{y})|_2.$$

Finally, by Poincaré's inequality there is $C > 0$ such that

$$\epsilon |y - \bar{y}|_2^2 + \epsilon C |(\beta + \epsilon I)(y) - (\beta + \epsilon I)(\bar{y})|_2^2$$
$$\le |x - \bar{x}|_2 \, |(\beta + \epsilon I)(y) - (\beta + \epsilon I)(\bar{y})|_2$$
$$\le \epsilon C |(\beta + \epsilon I)(y) - (\beta + \epsilon I)(\bar{y})|_2^2 + \frac{1}{\epsilon C} |x - \bar{x}|_2^2$$

and therefore

$$|y - \bar{y}|_2^2 \le \frac{1}{\epsilon^2 C} |x - \bar{x}|_2^2,$$

which proves that J_ϵ is Lipschitz continuous in L^2 as claimed, and (i) is proved.

(ii): Let $x \in L^p$. From (2.31) with $\bar{y} = 0$ we have

$$y - \epsilon \Delta (\beta + \epsilon I) y = x.$$

Multiplying both sides of the identity above by $y^{p-1}(1 + \lambda y^{p-2})^{-1}$, $\lambda > 0$, and integrating over $\mathcal{O}$ we get

$$\int_{\mathcal{O}} \frac{y^p}{1 + \lambda y^{p-2}} \, d\xi \le \int_{\mathcal{O}} \frac{y^{p-1} x}{1 + \lambda y^{p-2}} \, d\xi.$$

Then letting $\lambda \to 0$ we find the estimate

$$|y|_p^p \le \int_{\mathcal{O}} y^{p-1} x \, d\xi \le |y|_p^{p-1} |x|_p.$$

Hence

$$|J_\epsilon(x)|_p \le |x|_p, \quad \forall \, x \in L^p,$$

which implies (2.30) because

$$\langle |x|^{p-2} x, F_\epsilon(x) \rangle = \frac{1}{\epsilon} \langle |x|^{p-2} x, x - J_\epsilon(x) \rangle \ge \frac{1}{\epsilon} (|x|_p^p - |J_\epsilon(x)|_p) \ge 0.$$

$\square$

We shall also need some identities.

Lemma 2.3.2 *For all $x \in H^{-1}$ and all $\epsilon > 0$ we have*

$$\langle F_\epsilon(x), x\rangle_{-1} = \langle(\beta + \epsilon I)(J_\epsilon(x)), J_\epsilon(x)\rangle_2 + \epsilon\|F_\epsilon(x)\|_{-1}^2 \tag{2.33}$$

and for all $x \in L^2$

$$\langle F_\epsilon(x), x\rangle_2 = \langle A(\beta + \epsilon I)(J_\epsilon(x)), J_\epsilon(x)\rangle_2 + \epsilon|F_\epsilon(x)|_2^2. \tag{2.34}$$

Proof To prove (2.33) we write

$$\langle F_\epsilon(x), x\rangle_{-1} = -\langle\Delta(\beta + \epsilon I)(J_\epsilon(x)), J_\epsilon(x)\rangle_{-1} + \langle F_\epsilon(x), x - J_\epsilon(x)\rangle_{-1}$$

$$= -\langle\Delta(\beta + \epsilon I)(J_\epsilon(x)), J_\epsilon(x)\rangle_{-1} + \epsilon\|F_\epsilon(x)\|_{-1}^2$$

$$= \langle(\beta + \epsilon I)(J_\epsilon(x)), J_\epsilon(x)\rangle_2 + \epsilon\|F_\epsilon(x)\|_{-1}^2.$$

The proof of (2.34) is analogous due to the fact that by (2.23) it follows that $\Delta(\beta + \epsilon I)(J_\epsilon(x)) \in L^2$ if $x \in L^2$ and so, by elliptic regularity, $(\beta + \epsilon I)(J_\epsilon(x)) \in H^2 \cap H_0^1$. $\qquad\square$

2.4 Convergence of $\{X_\epsilon\}$

We are going to show that $\{X_\epsilon\}$ is a Cauchy sequence in the space $L_W^2(\Omega; C([0, T]; H^{-1}))$. The idea of the proof is typical for problems with monotone nonlinearities. It is based on a suitable a priori estimate for

$$\mathbb{E}\int_0^t \|F_\epsilon(X_\epsilon(s))\|_{-1}^2\, ds$$

and the following identity.

Lemma 2.4.1 *For all $x, y \in H^{-1}$ and $\epsilon, \eta > 0$ we have*

$$\langle F_\epsilon(x) - F_\eta(y), x - y\rangle_{-1} = \langle(\beta + \epsilon I)(J_\epsilon(x)) - (\beta + \epsilon I)(J_\eta(y)), J_\epsilon(x) - J_\eta(y)\rangle_2$$

$$+(\epsilon - \eta)\langle J_\eta(y), J_\epsilon(x) - J_\eta(y)\rangle_2 \tag{2.35}$$

$$+\langle F_\epsilon(x) - F_\eta(y), \epsilon F_\epsilon(x) - \eta F_\eta(y)\rangle_{-1}.$$

Proof We have

$$\langle F_\epsilon(x) - F_\eta(y), x - y\rangle_{-1} = \langle F_\epsilon(x) - F_\eta(y), J_\epsilon(x) - J_\eta(y)\rangle_{-1} \tag{2.36}$$

$$+\langle F_\epsilon(x) - F_\eta(y), \epsilon F_\epsilon(x) - \eta F_\eta(y)\rangle_{-1}.$$

On the other hand,

$$\langle F_\epsilon(x) - F_\eta(y), J_\epsilon(x) - J_\eta(y)\rangle_{-1}$$
$$= -\langle \Delta(\beta + \epsilon I)J_\epsilon(x) - \Delta(\beta + \eta I)J_\eta(y), J_\epsilon(x) - J_\eta(y)\rangle_{-1}$$

$$= -\langle \Delta(\beta + \epsilon I)J_\epsilon(x) - \Delta(\beta + \epsilon I)J_\eta(y), J_\epsilon(x) - J_\eta(y)\rangle_{-1}$$

$$-(\epsilon - \eta)\langle \Delta J_\eta(y), J_\epsilon(x) - J_\eta(y)\rangle_{-1}.$$

Substituting this in (2.36) yields (2.35).   □

We need now a few estimates on X_ϵ.

2.4.1 Estimates for $\|X_\epsilon(t)\|^2_{-1}$

Lemma 2.4.2 *Assume that Hypothesis 1(i), (ii), (iv) holds. Let $x \in H^{-1}$, $\epsilon > 0$ and let X_ϵ be the solution to (2.20). Then for any $T > 0$ the following inequality holds*

$$\mathbb{E}\|X_\epsilon(t)\|^2_{-1} + 2\mathbb{E}\int_0^t \langle J_\epsilon(X_\epsilon(s)), (\beta + \epsilon I)(J_\epsilon(X_\epsilon(s)))\rangle_2 \, ds$$

$$+ 2\epsilon\mathbb{E}\int_0^t \|F_\epsilon(X_\epsilon(s))\|^2_{-1} \, ds \tag{2.37}$$

$$\leq e^{2K_1^2 t}(\|x\|^2_{-1} + 2\sigma_1^2 T), \quad t \in [0, T].$$

Moreover,

$$\mathbb{E} \sup_{t \in [0,T]} \|X_\epsilon(t)\|^2_{-1} \leq (2\|x\|^2_{-1} + 4(c_1^2 + 1)\sigma_1^2 T) \, e^{4(c_1^2 + 1)K_1^2 T}, \tag{2.38}$$

where σ_1, K_1 are as in (2.6) and c_1 as in (1.23).

Proof By Itô's formula in H^{-1} we have

$$d\|X_\epsilon(t)\|^2_{-1} = -2\langle X_\epsilon(t), F_\epsilon(X_\epsilon(t))\rangle_{-1} \, dt + \mathrm{Tr}\,[\sigma(X_\epsilon(t))\sigma(X_\epsilon(t))^*] \, dt$$

$$+ 2\langle X_\epsilon(t), \sigma(X_\epsilon(t))dW(t)\rangle_{-1}. \tag{2.39}$$

Now, taking into account (2.33), yields

$$d\|X_\epsilon(t)\|_{-1}^2 + 2\langle(\beta + \epsilon I)(J_\epsilon(X_\epsilon(t))), J_\epsilon(X_\epsilon(t))\rangle_2\, dt + 2\epsilon\|F_\epsilon(X_\epsilon(t))\|_{-1}^2\, dt$$

$$= \text{Tr}\,[\sigma(X_\epsilon(t))\sigma(X_\epsilon(t))^*]\, dt + 2\langle X_\epsilon(t), \sigma(X_\epsilon(t))dW(t)\rangle_{-1}.$$

(2.40)

Then, integrating with respect to t, we obtain

$$\|X_\epsilon(t)\|_{-1}^2 + 2\int_0^t \langle J_\epsilon(X_\epsilon(s)), (\beta + \epsilon I)(J_\epsilon(X_\epsilon(s)))\rangle_2\, ds$$

$$+ 2\epsilon \int_0^t \|F_\epsilon(X_\epsilon(s))\|_{-1}^2\, ds$$

$$= \|x\|_{-1}^2 + 2\int_0^t \langle X_\epsilon(s), \sigma(X_\epsilon(s))dW(s)\rangle_{-1}$$

(2.41)

$$+ \int_0^t \text{Tr}\,[\sigma(X_\epsilon(s))\sigma(X_\epsilon(s))^*]\, ds.$$

Taking expectation and differentiating with respect to t, yields

$$\frac{d}{dt}\,\mathbb{E}\|X_\epsilon(t)\|_{-1}^2 + 2\mathbb{E}\langle J_\epsilon(X_\epsilon(t)), (\beta + \epsilon I)(J_\epsilon(X_\epsilon(t)))\rangle_2$$

(2.42)

$$+ 2\epsilon\mathbb{E}\|F_\epsilon(X_\epsilon(t))\|_{-1}^2 = \mathbb{E}\text{Tr}\,[\sigma(X_\epsilon(t))\sigma(X_\epsilon(t))^*].$$

Finally, recalling (2.6), yields

$$\frac{d}{dt}\,\mathbb{E}\|X_\epsilon(t)\|_{-1}^2 + 2\mathbb{E}\langle J_\epsilon(X_\epsilon(t)), (\beta + \epsilon I)(J_\epsilon(X_\epsilon(t)))\rangle_2$$

(2.43)

$$+ 2\epsilon\mathbb{E}\|F_\epsilon(X_\epsilon(t))\|_{-1}^2 \leq 2\sigma_1^2 + 2K_1^2\mathbb{E}\|X_\epsilon(t)\|_{-1}^2,$$

and (2.37) follows by a standard comparison result.

Let us now prove (2.38). By (2.41) and (2.6) we have for $t \in [0, T]$

$$\sup_{r\in[0,t]} \|X_\epsilon(r)\|_{-1}^2 \leq \|x\|_{-1}^2 + 2\sup_{r\in[0,t]} \int_0^r \langle X_\epsilon(s), \sigma(X_\epsilon(s))dW(s)\rangle_{-1}$$

(2.44)

$$+ 2\sigma_1^2 t + 2K_1^2 \int_0^t \|X_\epsilon(s)\|_{-1}^2 ds.$$

Now by the Burkholder–Davis–Gundy inequality with $p = 1$ (see (1.23)) and (2.6) we have

$$\mathbb{E} \sup_{r \in [0,t]} \left| \int_0^r \langle X_\epsilon(s), \sigma(X_\epsilon(s)) dW(s) \rangle_{-1} \right|$$

$$\leq c_1 \mathbb{E} \left(\int_0^t |\sigma(X_\epsilon(s))^* X_\epsilon(s)|_{-1}^2 \, ds \right)^{1/2}$$

$$\leq \sqrt{2} c_1 \mathbb{E} \left[\left(\int_0^t (\sigma_1^2 \|X_\epsilon(s)\|_{-1}^2 + K_1^2 \|X_\epsilon(s)\|_{-1}^4) ds \right)^{1/2} \right].$$

We have therefore

$$\mathbb{E} \sup_{r \in [0,t]} \left| \int_0^r \langle X_\epsilon(s), \sigma(X_\epsilon(s)) dW(s) \rangle_{-1} \right|$$

$$\leq \sqrt{2} \, c_1 \mathbb{E} \left[\left(\sup_{r \in [0,t]} \|X_\epsilon(r)\|_{-1}^2 \int_0^t (\sigma_1^2 + K_1^2 \|X_\epsilon(r)\|_{-1}^2) ds \right)^{1/2} \right] \quad (2.45)$$

$$\leq \frac{1}{4} \mathbb{E} \sup_{r \in [0,t]} \|X_\epsilon(r)\|_{-1}^2 + 2c_1^2 \mathbb{E} \int_0^t (\sigma_1^2 + K_1^2 \|X_\epsilon(s)\|_{-1}^2) ds.$$

Now taking into account (2.44), yields

$$\mathbb{E} \sup_{r \in [0,t]} \|X_\epsilon(r)\|_{-1}^2 \leq \|x\|_{-1}^2 + \frac{1}{2} \mathbb{E} \sup_{r \in [0,t]} \|X_\epsilon(r)\|_{-1}^2$$

$$+ 2(c_1^2 + 1) \left(\sigma_1^2 t + K_1^2 \mathbb{E} \int_0^t \sup_{r \in [0,s]} \|X_\epsilon(s)\|_{-1}^2 \, ds \right),$$

and therefore (2.38) follows by Gronwall's Lemma. $\square$

2.4.2 *Estimates for* $\mathbb{E} \int_0^t \|F_\epsilon(X_\epsilon(s))\|_{-1}^2 \, ds$

We start by estimating $|X_\epsilon(t)|_2^2$ using Itô's formula in L^2.

Lemma 2.4.3 *Assume that Hypothesis 1 holds. Then for each $x \in L^2$, $T > 0$, and each $\epsilon > 0$ we have*

$$\mathbb{E}|X_\epsilon(t)|_2^2 + 2\mathbb{E}\int_0^t \langle A(\beta + \epsilon I)(J_\epsilon(X_\epsilon(s))), J_\epsilon(X_\epsilon(s))\rangle_2 \, ds$$

(2.46)

$$\leq (|x|_2^2 + 2\sigma_2^2 T)e^{2K_2^2 T}, \quad \forall \, t \in [0, T].$$

Proof Applying Itô's formula to $|X_\epsilon(t)|_2^2$, yields

$$d|X_\epsilon(t)|_2^2 = -2\langle F_\epsilon(X_\epsilon(t)), X_\epsilon(t)\rangle_2 dt + \|\sigma(X_\epsilon(t))\|_{\mathscr{L}_2(H^{-1}, L^2)}^2 \, dt$$

(2.47)

$$+2\langle X_\epsilon(t)), \sigma(X_\epsilon(t))dW(t)\rangle_2.$$

Taking into account (2.34), yields

$$d|X_\epsilon(t)|_2^2 + 2\langle A(\beta + \epsilon I)(J_\epsilon(X_\epsilon(t))), J_\epsilon(X_\epsilon(t))\rangle_2 + 2\epsilon\,|F_\epsilon(X_\epsilon(t))|_2^2$$

(2.48)

$$= \|\sigma(X_\epsilon(t))\|_{\mathscr{L}_2(H^{-1}, L^2)}^2 \, dt + 2\langle X_\epsilon(t)), \sigma(X_\epsilon(t))dW(t)\rangle_2.$$

Integrating with respect to t and taking expectation, yields

$$\mathbb{E}|X_\epsilon(t)|_2^2 + 2\mathbb{E}\int_0^t \langle A(\beta + \epsilon I)(J_\epsilon(X_\epsilon(s))), J_\epsilon(X_\epsilon(s))\rangle_2 \, ds$$

(2.49)

$$+2\epsilon\mathbb{E}\int_0^t |F_\epsilon(X_\epsilon(s))|_2^2 \, ds = \mathbb{E}\int_0^t \|\sigma(X_\epsilon(s))\|_{\mathscr{L}_2(H^{-1}, L^2)}^2 \, ds.$$

Now differentiating with respect to t and taking into account (2.8) we get

$$\frac{d}{dt}\,\mathbb{E}|X_\epsilon(t)|_2^2 + 2\mathbb{E}\langle (A(\beta + \epsilon I)(J_\epsilon(X_\epsilon(t))), J_\epsilon(X_\epsilon(t))\rangle_2$$

$$+2\epsilon\mathbb{E}\,|F_\epsilon(X_\epsilon(t))|_2^2 = \mathbb{E}\|\sigma(X_\epsilon(t))\|_{\mathscr{L}_2(H^{-1}, L^2)}^2 \leq 2\sigma_2^2 + 2K_2^2\,\mathbb{E}|X_\epsilon(t)|_2^2$$

(2.50)

and the conclusion follows.   □

We are now ready to estimate

$$\mathbb{E}\int_0^t \|F_\epsilon(X_\epsilon(s))\|_{-1}^2 \, ds.$$

Proposition 2.4.4 *Assume that Hypothesis 1 holds. Then for each $x \in L^2$ and each $\epsilon \in (0, 1)$ we have*

$$\mathbb{E}\int_0^t \|F_\epsilon(X_\epsilon(s))\|_{-1}^2 \, ds \leq \frac{K+1}{2}(|x|_2^2 + 2\sigma_2^2 T)e^{2K_2^2 T}, \quad \forall \, t \in [0, T]. \qquad (2.51)$$

Proof Note first that

$$\|F_\epsilon(x)\|_{-1}^2 = \langle A(\beta + \epsilon I)(J_\epsilon(x)), (\beta + \epsilon I)(J_\epsilon(x))\rangle_2$$

$$\leq (K + \epsilon)\langle A(\beta + \epsilon I)(J_\epsilon(x)), J_\epsilon(x)\rangle_2,$$

(2.52)

because β is K-Lipschitz. Now the conclusion follows from (2.46) choosing $\epsilon < K$.

$\square$

2.4.3 Additional Estimates in L^p

In this section we shall prove, under a further assumption, Hypothesis 2 below, some estimates in L^p norm, $p \geq 2$, for the solution to (2.1). They will be used in Chap. 3 below.

Hypothesis 2 *There exists $\sigma_3 > 0$ and $K_3 > 0$ such that*

$$\sum_{k=1}^\infty [(\sigma(x)f_k)(\xi)]^2 \leq \sigma_3^2 + K_3^2 |x(\xi)|^2, \quad \forall x \in L^2, \xi \in \mathcal{O},$$

(2.53)

where $\{e_k\}$ is an orthonormal basis on L^2 of eigenfunctions of $A = -\Delta = H_0^1 \cap H^2$.

Example 2.4.5 Let us first consider the additive noise seen in Example 2.1.1 where $\sigma(x) = A^{-\gamma/2}$. Then we have

$$\sum_{k=1}^\infty [(\sigma(x)f_k)(\xi)]^2 = \sum_{k=1}^\infty \alpha_k^{-\gamma} [f_k(\xi)]^2.$$

Recalling (1.20) we find

$$\sum_{k=1}^\infty [(\sigma(x)f_k)(\xi)]^2 \leq b_\mathcal{O} \sum_{k=1}^\infty \mu_k \alpha_k^{d-\gamma}.$$

So, by (1.19) Hypothesis 2 is fulfilled provided

$$\gamma > \frac{3}{2}d.$$

(2.54)

Now we consider the linear noise from Example 2.1.2 by choosing

$$\sigma(x)h = KxA^{-\gamma/2}h, \quad h \in H^{-1},$$

and try to check (2.53) with $\sigma_3 = 0$. We have

$$\sum_{k=1}^{\infty} [(\sigma(x)f_k)(\xi)]^2 = K^2 \sum_{k=1}^{\infty} \alpha_k^{-\gamma} (\kappa(\xi))^2 \, |x(\xi)|^2.$$

Now recalling (1.20) we find

$$\sum_{k=1}^{\infty} [(\sigma(x)f_k)(\xi)]^2 \le K^2 b_{\mathcal{O}}^2 \sum_{k=1}^{\infty} \alpha_k^{d-\gamma} \, |x(\xi)|^2.$$

Consequently by (1.19), Hypothesis 2 is fulfilled provided

$$\gamma > \frac{3}{2} d, \tag{2.55}$$

which is the same condition we need for Hypothesis 1 (see Example 2.1.2).   □

Proposition 2.4.6 *Assume that Hypotheses 1 and 2 are fulfilled. Let $p \ge 2$, $x \in L^p$ and let X_ϵ be the solution to (2.20). Then X_ϵ is an L^p-valued adapted continuous process and there exists a constant $M_{1,p} > 0$ such that*

$$\sup_{t \in [0,T]} \mathbb{E}|X_\epsilon(t)|_p^p \le e^{M_{1,p}t}(|x|_p^p + M_{1,p}T). \tag{2.56}$$

Proof We are going to apply Proposition 1.2.3. To this end we shall prove the following:

Claim $X_\epsilon \in L^p(\Omega \times [0,T]; L^p)$.

Suppose we have proved this claim, then by Hypotheses 1 and 2 obviously all assumptions in Proposition 1.2.3 are fulfilled with $\tau := T, f := -F_\epsilon(X_\epsilon)$ and $g_k := \sigma(X_\epsilon)f_k$, $k \in \mathbb{N}$.

The Claim follows from the following stronger fact:

$$X_\epsilon \in L^\infty_W([0,T]; L^p(\Omega, L^p)) \tag{2.57}$$

To prove (2.57) for $R > 0$, $\alpha > 0$ consider the set

$$\mathscr{K}_R := \{X \in L^\infty_W([0,T]; L^p(\Omega, L^p)) : e^{-p\alpha t} \, \mathbb{E}|X(t)|_p^p \le R^p, \ t \in [0,T]\}$$

Since by (2.20) X_ϵ is a fixed point of the map Γ defined by

$$\Gamma(X) := e^{-\frac{t}{\epsilon}}x + \frac{1}{\epsilon}\int_0^t e^{-\frac{t-s}{\epsilon}} J_\epsilon(X(s))ds + \int_0^t e^{-\frac{t-s}{\epsilon}} \sigma(X(s))dW(s),$$

obtained by iteration in $C_W([0, T]; L^2(\Omega; H \cap L^2))$, it suffices to show that Γ leaves the set $\mathcal{K}_R$ invariant for $R > 0$ large enough. By (2.29) we have for every $X \in \mathcal{K}_R$

$$\left(e^{-p\alpha t} \, \mathbb{E} \left| e^{-\frac{t}{\epsilon}} x + \frac{1}{\epsilon} \int_0^t e^{-\frac{t-s}{\epsilon}} J_\epsilon(X(s)) ds \right|_p^p \right)^{\frac{1}{p}}$$

$$\leq e^{-(\frac{1}{\epsilon}+\alpha)t} |x|_p + \frac{e^{-\alpha t}}{\epsilon} \int_0^t e^{-\frac{t-s}{\epsilon}} \left(\mathbb{E}|X(s)|_p^p \right)^{\frac{1}{p}} ds \qquad (2.58)$$

$$\leq e^{-(\frac{1}{\epsilon}+\alpha)t)} |x|_p + \frac{R}{1+\alpha\epsilon} .$$

Now we set

$$Y(t) := \int_0^t e^{-\frac{t-s}{\epsilon}} \sigma(X(s)) dW(s), \quad t \geq 0.$$

We have

$$\begin{cases} dY + \frac{1}{\epsilon} Y \, dt = \sigma(X) \, dW, & t \geq 0, \\ Y(0) = 0. \end{cases}$$

Equivalently

$$d(e^{\frac{t}{\epsilon}} Y(t)) = e^{\frac{t}{\epsilon}} \sigma(X(t)) \, dW(t), \quad t > 0, \quad Y(0) = 0.$$

By Proposition 1.2.3 it follows that $e^{\frac{t}{\epsilon}} Y$ is an L^p-valued $(\mathscr{F}_t)$-adapted continuous process on $[0, \infty)$ and

$$\mathbb{E}|e^{\frac{t}{\epsilon}} Y(t)|_p^p \leq \frac{1}{2} p(p-1) \mathbb{E} \int_0^t \int_{\mathscr{O}} |e^{\frac{s}{\epsilon}} Y(s)|^{p-2} \sum_{k=1}^{\infty} |e^{\frac{s}{\epsilon}} \sigma(X(s)) f_k|^2 \, d\xi \, ds.$$

This yields via Hypothesis 2

$$\mathbb{E}|e^{\frac{t}{\epsilon}} Y(t)|_p^p \leq \frac{1}{2} p(p-1) \mathbb{E} \int_0^t \int_{\mathscr{O}} |e^{\frac{s}{\epsilon}} Y(s)|^{p-2} \, e^{\frac{2s}{\epsilon}} (\sigma_3^2 + K_3^2 |X(s)|^2) \, d\xi \, ds,$$

which by Hölder's and the Hausdorff–Young inequality implies that

$$\mathbb{E}|e^{\frac{t}{\epsilon}} Y(t)|_p^p \leq \frac{1}{2} (p-1)(p-2) \mathbb{E} \int_0^t |e^{\frac{s}{\epsilon}} Y(s)|_p^p \, ds$$

$$+ C \mathbb{E} \int_0^t e^{\frac{ps}{\epsilon}} (1 + |X(s)|_p^p) \, ds, \quad \forall \, t \in [0, T]$$

and therefore

$$e^{-p\alpha t}\,\mathbb{E}|Y(t)|_p^p \leq C_1 e^{-(\alpha+\frac{1}{\epsilon})pt}\,\mathbb{E}\int_0^t e^{\frac{ps}{\epsilon}}\,(1+|X(s)|_p^p)\,ds$$

$$\leq \tfrac{C_1\epsilon(1+R^p)}{p(1+\epsilon\alpha)} \quad \forall\, t \in [0,T].$$

Then by formula (2.58) we infer that, for α large enough and $R > 2|x|_p$, Γ leaves $\mathcal{K}_R$ invariant which proves (2.57).

Now we know by Proposition 1.2.3 that X_ϵ is an L^p-valued adapted continuous process and for all $t \in [0,T]$

$$\mathbb{E}|X_\epsilon(t)|_p^p = -p\mathbb{E}\int_0^t \langle |X_\epsilon(s)|^{p-2} X_\epsilon(s), F_\epsilon(X_\epsilon(s))\rangle\, ds$$

$$+|x|_p^p + p(p-1)\sum_{k=1}^\infty \mathbb{E}\int_0^t \int_{\mathcal{O}} |X_\epsilon(s)|^{p-2}\,|\sigma(X_\epsilon(s))f_k|^2 d\xi\, ds.$$

$$\tag{2.59}$$

Now from (2.30) we deduce

$$\mathbb{E}|X_\epsilon(t)|_p^p \leq |x|_p^p + p(p-1)\sum_{k=1}^\infty \mathbb{E}\int_0^t \int_{\mathcal{O}} |X_\epsilon(s)|^{p-2}\,|\sigma(X_\epsilon(s))f_k|^2 d\xi\, ds,$$

which by Hypothesis 2 implies

$$\mathbb{E}|X_\epsilon(t)|_p^p \leq |x|_p^p + p(p-1)\mathbb{E}\int_0^t \int_{\mathcal{O}} (|X_\epsilon(s)|^{p-2}\,\sigma_3^2 + K_3^2|X_\epsilon(s)|^p)\,d\xi\, ds.$$

Let us note that there exists $M_{1,p} > 0$ such that

$$p(p-1)r^{p-2}(\sigma_3^2 + K_3^2 r^2) \leq M_{1,p}\,(1 + r^p), \quad \forall\, r > 0.$$

Consequently from (2.59) we deduce

$$\mathbb{E}|X_\epsilon(t)|_p^p \leq |x|_p^p + M_{1,p}\mathbb{E}\int_0^t \int_{\mathcal{O}} (|X_\epsilon(s)|^p + 1)\,d\xi\, ds,$$

so that

$$\mathbb{E}|X_\epsilon(t)|_p^p \leq |x|_p^p + M_{1,p}\mathbb{E}\int_0^t (|X_\epsilon(s)|_p^p + 1)\,d\xi\, ds.$$

Now the conclusion follows by Gronwall's lemma.   □

2.5 The Solution to Problem (2.1)

Theorem 2.5.1 *Assume that Hypothesis 1 is fulfilled. Then for each $T > 0$ and all $x \in L^2$ problem (2.1) has a unique strong solution. Moreover, for $x \in H^{-1}$ problem (2.1) has a unique generalized solution. Furthermore, if $x \in L^p$, $p \geq 2$, then*

$$\sup_{t \in [0,T]} \mathbb{E}|X(t)|_p^p \leq e^{M_{1,p}T}(|x|_p^p + M_{1,p}\, T). \tag{2.60}$$

Proof Step 1. For each $x \in L^2$ the sequence $\{X_\epsilon\}$ (where X_ϵ is the solution of (2.20)) is Cauchy in $C_W([0,T]; L^2(\Omega; H^{-1}))$.

Let $\epsilon, \eta > 0$. Using Itô's formula for $\|X_\epsilon(t) - X_\eta(t)\|_{-1}^2$, yields

$$d\|X_\epsilon(t) - X_\eta(t)\|_{-1}^2 = -2\langle X_\epsilon(t) - X_\eta(t), F_\epsilon(X_\epsilon(t)) - F_\eta(X_\eta(t))\rangle_{-1}\, dt$$

$$+ 2\langle X_\epsilon(t) - X_\eta(t), (\sigma(X_\epsilon(t)) - \sigma(X_\eta(t))dW(t)\rangle_{-1}$$

$$+ \|\sigma(X_\epsilon(t)) - \sigma(X_\eta(t))\|_{\mathscr{L}_2(H^{-1})}^2\, dt. \tag{2.61}$$

Recalling identity (2.35) we obtain that

$$d\|X_\epsilon(t) - X_\eta(t)\|_{-1}^2 + 2\langle(\beta + \epsilon I)J_\epsilon(X_\epsilon(t))$$

$$-(\beta + \epsilon I)J_\eta(X_\eta(t)), J_\epsilon(X_\epsilon(t)) - J_\eta(X_\eta(t))\rangle_2\, dt$$

$$= -2(\epsilon - \eta)\langle J_\eta(X_\eta(t)), J_\epsilon(X_\epsilon(t)) - J_\eta(X_\eta(t))\rangle_2\, dt$$

$$-2\langle F_\epsilon(X_\epsilon(t)) - F_\eta(X_\eta(t)), \epsilon F_\epsilon(X_\epsilon(t)) - \eta F_\eta(X_\eta(t))\rangle_{-1}\, dt$$

$$+2\langle X_\epsilon(t) - X_\eta(t), (\sigma(X_\epsilon(t)) - \sigma(X_\eta(t))dW(t)\rangle_{-1}$$

$$+\|\sigma(X_\epsilon(t)) - \sigma(X_\eta(t))\|_{\mathscr{L}_2(H^{-1})}^2\, dt. \tag{2.62}$$

Now, integrating with respect to t and taking into account (2.5), yields

$$\|X_\epsilon(t) - X_\eta(t)\|_{-1}^2 + 2\int_0^t \langle(\beta + \epsilon I)J_\epsilon(X_\epsilon(s))$$

$$-(\beta + \epsilon I)J_\eta(X_\eta(s)), J_\epsilon(X_\epsilon(s)) - J_\eta(X_\eta(s))\rangle_2\, ds$$

$$\leq 2\int_0^t \langle X_\epsilon(s) - X_\eta(s), (\sigma(X_\epsilon(s)) - \sigma(X_\eta(s))dW(s)\rangle_{-1}$$

$$+2(\epsilon + \eta)\int_0^t |X_\eta(s)|_2(|X_\epsilon(s)|_2 + |X_\eta(s)|_2)ds$$

$$+2(\epsilon + \eta) \int_0^t (\|F_\epsilon(X_\epsilon(s))\|_{-1}^2 + \|F_\eta(X_\eta(s))\|_{-1}^2) ds$$

$$+K_1 \int_0^t \|X_\epsilon(s) - X_\eta(s)\|_{-1}^2 ds. \tag{2.63}$$

Taking expectation and taking into account (2.46) and (2.51), yields

$$\mathbb{E}\|X_\epsilon(t) - X_\eta(t)\|_{-1}^2 + 2\mathbb{E} \int_0^t \langle (\beta + \epsilon I)J_\epsilon(X_\epsilon(s))$$

$$-(\beta + \epsilon I)J_\eta(X_\eta(s)), J_\epsilon(X_\epsilon(s)) - J_\eta(X_\eta(s)) \rangle_2 ds$$

$$\leq 2T(K+3)(\epsilon + \eta)(|x|_2^2 + 2\sigma_2^2 T)e^{2K_2^2 T}$$

$$+K_1 \int_0^t \mathbb{E}\|X_\epsilon(s) - X_\eta(s)\|_{-1}^2 ds. \tag{2.64}$$

Finally, applying a standard comparison result (or Gronwall's lemma), yields

$$\mathbb{E}\|X_\epsilon(t) - X_\eta(t)\|_{-1}^2 + 2\mathbb{E} \int_0^t \langle (\beta + \epsilon I)J_\epsilon(X_\epsilon(s))$$

$$-(\beta + \epsilon I)J_\eta(X_\eta(s)), J_\epsilon(X_\epsilon(s)) - J_\eta(X_\eta(s)) \rangle_2 ds$$

$$\leq 2T(K+3)(\epsilon + \eta)(|x|_2^2 + 2\sigma_2^2 T)e^{2K_2^2 T}e^{K_1 T}. \tag{2.65}$$

Therefore, the sequence $\{X_\epsilon\}$ is Cauchy in $C_W([0, T]; L^2(\Omega; H^{-1}))$ as claimed.
Step 2. For each $x \in L^2$ the sequence $\{X_\epsilon\}$ is Cauchy in $L_W^2(\Omega; C([0, T]; H^{-1}))$.
    We first note that by (2.63), (2.46) and (2.51) we deduce

$$\mathbb{E} \sup_{t \in [0,T]} \|X_\epsilon(t) - X_\eta(t)\|_{-1}^2$$

$$\leq 2\mathbb{E} \sup_{t \in [0,T]} \int_0^t \langle X_\epsilon(s) - X_\eta(s), (\sigma(X_\epsilon(s)) - \sigma(X_\eta(s))dW(s) \rangle_{-1}$$

$$+2T(K+3)(\epsilon + \eta)(|x|_2^2 + 2\sigma_2^2 T)e^{2K_2^2 T} + K_1\mathbb{E} \int_0^T \|X_\epsilon(s) - X_\eta(s)\|_{-1}^2 ds. \tag{2.66}$$

Let us now estimate

$$\Sigma := \mathbb{E} \left| \sup_{t \in [0,T]} \int_0^t \langle X_\epsilon(s) - X_\eta(s)), (\sigma(X_\epsilon(s)) - \sigma(X_\eta(s)) dW(s) \rangle_{-1} ds \right|.$$

By using the Burkholder–Davis–Gundy inequality (1.23) we obtain

$$\Sigma \le c_1 \mathbb{E}\left[\left(\int_0^T |(\sigma(X_\epsilon(s)) - \sigma(X_\eta(s)))^* (X_\epsilon(s) - X_\eta(s))|^2_{-1} \, ds\right)^{1/2}\right].$$

$$\le c_1 K_1 \mathbb{E}\left[\left(\int_0^T \|X_\epsilon(s) - X_\eta(s)\|^4_{-1} ds\right)^{1/2}\right]$$

$$\le c_1 K_1 \mathbb{E}\left[\left(\sup_{t\in[0,T]} \|X_\epsilon(t) - X_\eta(t)\|^2_{-1} \int_0^T \|X_\epsilon(s) - X_\eta(s)\|^2_{-1}) ds\right)^{1/2}\right]$$

$$\le \tfrac{1}{4} \mathbb{E}\sup_{t\in[0,T]} \|X_\epsilon(t) - X_\eta(t)\|^2_{-1} + c_1^2 K_1^2 \mathbb{E}\int_0^T \|X_\epsilon(s) - X_\eta(s)\|^2_{-1}) ds.$$
(2.67)

Now, substituting in (2.66) we have

$$\frac{1}{2} \mathbb{E}\sup_{t\in[0,T]} \|X_\epsilon(t) - X_\eta(t)\|^2_{-1} \le (2c_1^2 K_1^2 + K_1)\mathbb{E}\int_0^T \|X_\epsilon(s) - X_\eta(s)\|^2_{-1} ds$$

$$+ T(K+3)(\epsilon + \eta)(|x|^2_2 + 2\sigma_2^2 T)e^{2K_2^2 T}.$$
(2.68)

Now Step 2 follows from Step 1.

Step 3. Existence of a strong solution.

Since $\{X_\epsilon\}$ is Cauchy in $L^2_W(\Omega; C([0,T]; H^{-1}))$ there exists $X \in L^2_W(\Omega; C([0,T]; H^{-1}))$ such that

$$\lim_{\epsilon\to 0} X_\epsilon = X \quad \text{in } L^2_W(\Omega; C([0,T]; H^{-1})).$$
(2.69)

It remains to show that $\beta(X) \in L^2(0,T; L^2(\Omega; H_0^1))$ and that (2.17) is fulfilled. We first note that

$$\lim_{\epsilon\to 0} J_\epsilon(X_\epsilon) = X \quad \text{in } L^2(0,T; L^2(\Omega; H^{-1}))$$
(2.70)

because,

$$\|X(t) - J_\epsilon(X(t))\|_{-1} \le \epsilon \|F_\epsilon(X_\epsilon(t))\|_{-1}$$

and in view of (2.51),

$$\mathbb{E}\int_0^T \|X_\epsilon(t) - J_\epsilon(X_\epsilon(t))\|^2_{-1} dt \le \epsilon \mathbb{E}\int_0^T \|F_\epsilon(X_\epsilon(t))\|^2_{-1} dt$$

$$\le \epsilon \frac{K+1}{2} (|x|^2_2 + 2\sigma_2^2 T)e^{2K_2^2 T}.$$
(2.71)

Moreover from (2.65) and (2.9) it follows that $\{\beta(J_\epsilon(X_\epsilon))\}$ is Cauchy in $L^2((0,T) \times \Omega \times \mathcal{O})$. So, there exists $Z \in L^2((0,T) \times \Omega \times \mathcal{O})$ such that

$$\lim_{\epsilon \to 0} \beta(J_\epsilon(X_\epsilon)) = Z \quad \text{in } L^2((0,T) \times \Omega \times \mathcal{O}). \tag{2.72}$$

On the other hand, by estimate (2.46) we know that $\{X_\epsilon\}$ is bounded in $L^2(\Omega \times [0,T] \times \mathcal{O})$ and therefore along a subsequence, again noted $\{\epsilon\}$, we have

$$\lim_{\epsilon \to 0} X_\epsilon = X \quad \text{weakly in } L^2(\Omega \times (0,T) \times \mathcal{O}). \tag{2.73}$$

Since

$$|J_\epsilon(X_\epsilon(t))|_2 \le |X_\epsilon(t)|_2, \quad \forall t \in [0,T],$$

we infer by (2.72) and (2.71) that

$$\lim_{\epsilon \to 0} J_\epsilon(X_\epsilon) = X \quad \text{weakly in } L^2(\Omega \times (0,T) \times \mathcal{O}).$$

Since the map $y \mapsto \beta(y)$ is maximal monotone on $L^2(\Omega \times (0,T) \times \mathcal{O})$, it is weakly-strongly closed (Proposition 1.2.9) and so $Z = \beta(X)$ a.e. in $\Omega \times (0,T) \times \mathcal{O}$.

Finally, since $\{(\beta + \epsilon I)(J_\epsilon(X_\epsilon))\}$ is bounded in $L^2([0,T]; L^2(\Omega, H_0^1))$ (because $F_\epsilon(X_\epsilon) = \Delta(\beta + \epsilon I)(J_\epsilon(X_\epsilon))$, is bounded in $L^2([0,T]; L^2(\Omega, H^{-1})))$, we can conclude that there exists a sequence $\{\epsilon_k\}$ convergent to 0 such that

$$\lim_{k \to \infty} (\beta + \epsilon_k I)(J_{\epsilon_k}(X_{\epsilon_k})) \to Z = \beta(X) \quad \text{weakly in } L^2([0,T] \times \Omega; H_0^1).$$

So, letting $k \to \infty$ in the identity

$$X_{\epsilon_k}(t) = x + \int_0^t \Delta(\beta + \epsilon_k I)(J_{\epsilon_k}(X_{\epsilon_k}(s)))ds + \sigma(X_{\epsilon_k}(t)))dW(t),$$

we infer that X is a strong solution to (2.1). Then (2.60) follows immediately from (2.56).

Step 4. Existence of a generalized solution.

Let $x \in H^{-1}$ and let $\{x_n\}$ be a sequence in L^2 convergent to x in H^{-1}. Denote by X_n the strong solution to (2.1) with x_n replacing x. Then for any $m, n \in \mathbb{N}$ we have by Itô's formula

$$d\|X_m(t) - X_n(t)\|_{-1}^2 + 2\langle F(X_m(t)) - F(X_n(t)), X_m(t) - X_n(t)\rangle_{-1}\, dt$$
$$+ 2\langle X_m(t) - X_n(t), (\sigma(X_m(t)) - \sigma(X_n(t)))dW(t)\rangle_{-1}$$
$$+ \|\sigma(X_m(t)) - \sigma(X_n(t))\|_{\mathscr{L}_2(H^{-1})}^2\, dt.$$

Integrating with respect to t it follows that

$$\|X_m(t) - X_n(t)\|_{-1}^2 \leq 2 \int_0^t \langle X_m(s) - X_n(s), (\sigma(X_m(s)) - \sigma(X_n(s)))dW(s)\rangle_{-1}$$

$$+ \int_0^t \|\sigma(X_m(s)) - \sigma(X_n(s))\|_{\mathscr{L}_2(H^{-1})}^2 \, ds.$$

Consequently

$$\mathbb{E} \sup_{t \in [0,T]} \|X_m(t) - X_n(t)\|_{-1}^2 \leq \|x_n - x_m\|_{-1}^2$$

$$+ 2\mathbb{E} \sup_{t \in [0,T]} \int_0^T \langle X_m(s) - X_n(s), (\sigma(X_m(s))$$

$$- \sigma(X_n(s)))dW(s)\rangle_{-1}$$

$$+ K_1 \mathbb{E} \int_0^T \|X_m(s)) - X_n(s))\|_{-1}^2 \, ds.$$

Using Burkholder–Davis–Gundy and Gronwall estimates as in the proof of Lemma 2.4.2, we conclude that the sequence $\{X_n\}$ is Cauchy in

$$L_W^2(\Omega; C([0, T]; H^{-1}))$$

and consequently it is convergent to a generalized solution of (2.1). □

2.6 Positivity of Solutions

Here we denote by L_+^2 the closed convex subset of L^2 of nonnegative functions,

$$L_+^2 = \{x \in L^2 : x(\xi) \geq 0, \text{ a.e. } \xi \in \mathcal{O}\}.$$

For any $x \in L^2$ we set

$$x^+ = \max\{x, 0\}, \quad x^- = \max\{-x, 0\}.$$

Let $x \in L^2$ and let $X(t, x)$ be the strong solution of (2.1). In this section we want to show that $X(t, x) \in L_+^2$ for all $x \in L_+^2$. For this we need, besides Hypothesis 1, the following

Hypothesis 3 *There exists $\sigma_3 > 0$, such that*

$$\sum_{k=1}^{\infty}[(\sigma(x)f_k)(\xi)]^2 \le \sigma_3|x(\xi)|^2, \quad \forall\, x \in L^2,\, \xi \in \mathcal{O}, \tag{2.74}$$

where $\{f_k\}$ is an orthonormal basis in H^{-1} introduced at the beginning of Sect. 2.1.

Obviously this assumption is never fulfilled for equations with additive noise. This is quite natural because the additive noise will eventually drive the system outside L_+^2. The situation is different in presence of multiplicative noise as the following example shows.

Example 2.6.1 (Linear Noise) We continue here Example 2.1.2 choosing

$$\sigma(x)h = KxA^{-\gamma/2}h, \quad h \in H^{-1}$$

and try to check (2.74). We recall that Hypothesis 1 is fulfilled provided $\gamma \ge \frac{3}{2}d$. Moreover, proceeding as in Example 2.4.5, we see that Hypothesis 3 is fulfilled as well provided

$$\gamma > \frac{3d}{2}. \tag{2.75}$$

Let $\mathscr{M}^+$ denote the set of all nonnegative Borel measures on $\mathcal{O}$ which are finite on compact subsets of $\mathcal{O}$.

Theorem 2.6.2 *Assume Hypotheses 1 and 3.*

 (i) Let $x \in L_+^2$ and let $X(\cdot, x)$ be the strong solution of (2.1). Then $X(t,x)(\omega) \in L_+^2$ for $dt \otimes \mathbb{P}$-a.e. $(t,\omega) \in [0,T] \times \Omega$.
 (ii) Let $x \in H^{-1} \cap \mathscr{M}^+$. Then $\mathbb{P}$-a.s. $X(t,x) \in H^{-1} \cap \mathscr{M}^+$ for all $t \in [0,T]$.

Proof We first note that by (2.73) it suffices to prove the assertion for the approximating solutions $\{X_\epsilon\}$ to Eq. (2.20) for initial condition $x \in L_+^2$. We recall that $t \to X_\epsilon(t)$ is continuous in L^2.

Define for $\delta \in (0,1)$

$$g_\delta(r) = \frac{r^2}{\delta + r}, \quad r \in (-\delta, \infty).$$

Then for $r \in (-\delta, \infty)$

$$g'_\delta(r) = 1 - \frac{\delta^2}{(\delta + r)^2} \in [0,1]$$

$$g''_\delta(r) = \frac{2\delta^2}{(\delta + r)^3}.$$

Furthermore, it is easy to check that

$$G_\delta(r) := g_\delta((r^-)^2), \quad r \in \mathbb{R},$$

is C^2 on $\mathbb{R}$ with

$$G_\delta'(r) = -2r^- \left(1 - \frac{\delta^2}{(\delta + (r^-)^2)^2} \right)$$

$$G_\delta''(r) = 2 + \left(\frac{8(r^-)^2}{\delta + (r^-)^2} - 2 \right) \frac{\delta^2}{(\delta + (r^-)^2)^2} \cdot$$

Hence $G_\delta(r) = G_\delta'(r) = G_\delta''(r) = 0$ for all $r \in [0, \infty)$, $|G_\delta'(r)| \leq 2r^-$ and $0 \leq G_\delta'' \leq 8$.
Now define $\varphi_\delta : L^2 \to \mathbb{R}$ by

$$\varphi_\delta(x) = \int_{\mathcal{O}} G_\delta(x) \, d\xi, \quad x \in L^2. \tag{2.76}$$

Then φ_δ is twice Gâteaux differentiable on L^2 and

$$D\varphi_\delta(x) = G_\delta'(x), \quad D^2\varphi_\delta(x)z = G_\delta''(x)z, \quad x, z \in L^2. \tag{2.77}$$

We claim that (see [21, Lemma 3.5])

$$\mathbb{E}\varphi_\delta(X_\epsilon(t)) + \mathbb{E} \int_0^t \langle F_\epsilon(X_\epsilon(s)), D\varphi_\delta(X_\epsilon(s)) \rangle \, ds$$

$$\tag{2.78}$$

$$= \varphi_\delta(x) + \frac{1}{2} \sum_{k=1}^\infty \mathbb{E} \int_0^t \int_{\mathcal{O}} G_\delta''(X_\epsilon(s)) \, |\sigma(X_\epsilon(s))f_k|^2 \, d\xi \, ds,$$

where X_ϵ are the solutions to (2.20).
To get (2.78) one approximates φ_δ by

$$\varphi_{\delta,\lambda}(x) = \varphi_\delta((1 + \lambda A_0)^{-1}x), \quad A_0 = -\Delta, \quad D(A_0) = H^2(\mathcal{O}) \cap H_0^1(\mathcal{O}), \quad \lambda > 0,$$

and takes into account that

$$D\varphi_{\delta,\lambda}(x) = -((1 + \lambda A_0)^{-1}x)^- D\varphi_\delta((1 + \lambda A_0)^{-1}x),$$

and

$$\langle D^2\varphi_{\delta,\lambda}(x)h, k \rangle = \langle D^2\varphi_\delta((1 + \lambda A_0)^{-1}x)((1 + \lambda A_0)^{-1}h, (1 + \lambda A_0)^{-1}k \rangle,$$

for $h, k, x \in L^2(\mathcal{O})$.

So, for $\lambda \to 0$ we have $\varphi_{\delta,\lambda}(x) \to \varphi_\delta(x)$ and $D\varphi_{\delta,\lambda}(x) \to D\varphi_\delta(x)$ in $L^2(\mathcal{O})$. Since $\varphi_{\delta,\lambda}$ is C^2 on L^2 we can use Itô's formula for $\varphi_{\delta,\lambda}$ and get

$$\mathbb{E}\, \varphi_{\delta,\lambda}(X_\epsilon)(t)) + \mathbb{E} \int_0^t \langle F_\epsilon(X(s)), D\varphi_{\delta,\lambda}(X_\epsilon)(s)) \rangle ds$$

$$= \varphi_{\delta,\lambda}(x)$$

$$+ \sum_{k=1}^\infty \int_0^t \int_{\mathcal{O}} G_\delta''((1 + \lambda A_0)^{-1} X_\epsilon(s)))$$

$$\times |\sigma((1 + \lambda A_0)^{-1} X_\epsilon(s))(1 + \lambda A_0)^{-1} f_k|^2 \, d\xi \, ds.$$

Then one gets (2.78) letting $\lambda \to 0$. Recalling that our initial condition is in L_+^2 and that $Y_\epsilon = J_\epsilon(X_\epsilon)$ we have by (2.77)–(2.78),

$$\mathbb{E}\, \varphi_\delta(X_\epsilon(t)) - \mathbb{E} \int_0^t \int_{\mathcal{O}} \Delta(\beta(Y_\epsilon(s)) + \epsilon J_\epsilon(Y_\epsilon(s))) \, G_\delta'(X_\epsilon(s)) \, d\xi \, ds$$

$$= \frac{1}{2} \sum_{k=1}^\infty \mathbb{E} \int_0^t \int_{\mathcal{O}} G_\delta''(X_\epsilon(s)) \, |\sigma(X_\epsilon(s)) f_k|^2 \, d\xi \, ds \qquad (2.79)$$

$$\leq 4\sigma^3 \, \mathbb{E} \int_0^t |X_\epsilon^-(s)|_2^2 \, ds,$$

where in the last step we used Hypothesis 3 and that $G_\delta'' = 0$ on $[0, \infty)$.

By Lebesgue's dominated convergence theorem we can take the limit $\delta \to 0$ and by (2.23) we obtain

$$\mathbb{E}|X_\epsilon^-(s)|_2^2 + \frac{2}{\epsilon} \mathbb{E} \int_0^t \int_{\mathcal{O}} (Y_\epsilon(s) - X_\epsilon(s)) X_\epsilon^-(s) \, d\xi \, ds \leq 4\sigma^3 \, \mathbb{E} \int_0^t |X_\epsilon^-(s)|_2^2 \, ds.$$
$$(2.80)$$

Moreover, we have by (2.23)

$$\int_{\mathcal{O}} X_\epsilon(t) \, Y_\epsilon^-(t) \, d\xi = \int_{\mathcal{O}} (Y_\epsilon(t) - \epsilon \Delta(\beta + \epsilon I) Y_\epsilon(t)) \, Y_\epsilon^-(t) \, d\xi$$

$$= - \int_{\mathcal{O}} |Y_\epsilon^-(t)|^2 \, d\xi - \epsilon \int_{\mathcal{O}} (\beta' + \epsilon) |\nabla Y_\epsilon^-(t)|^2 \, d\xi$$

$$\leq - \int_{\mathcal{O}} |Y_\epsilon^-(t)|^2 \, d\xi.$$

Hence

$$-|Y_\epsilon^-(t)|_2^2 \geq \int_{\mathcal{O}} (X_\epsilon^+(t) - X_\epsilon^-(t)) Y_\epsilon^-(t) \, d\xi \geq - \int_{\mathcal{O}} X_\epsilon^-(t) \, Y_\epsilon^-(t) \, d\xi$$

and therefore

$$|Y_\epsilon^-(t)|_2^2 \leq |X_\epsilon^-(t)|_2 \, |Y_\epsilon^-(t)|_2.$$

Hence $|Y_\epsilon^-(t)|_2 \leq |X_\epsilon^-(t)|_2$ and so

$$\int_{\mathcal{O}} Y_\epsilon^-(t) X_\epsilon^-(t) \, d\xi \leq |X_\epsilon^-(t)|_2 \, |Y_\epsilon^-(t)|_2 \leq |X_\epsilon^-(t)|_2^2 = -\int_{\mathcal{O}} X_\epsilon(t) X_\epsilon^-(t) \, d\xi.$$

(2.81)

Inserting the latter into (2.80) and taking into account that $Y_\epsilon X_\epsilon^- \geq -Y_\epsilon^- X_\epsilon^-$, we see that the second term in the right hand side of (2.80) is positive. So, by Gronwall's lemma it follows that $\mathbb{E}\,|X_\epsilon^-(t)|_2^2 = 0$, for a.e. $t \geq 0$ i.e., $X_\epsilon^- = 0$ and therefore $X_\epsilon \geq 0$ a.e. on $(0, T) \times \mathcal{O}$.

(ii) If $x \in L_+^2$ the assertion follows immediately, from the continuity of $t \to X(t, x)$ in H^{-1}.

If $x \in H^{-1} \cap \mathcal{M}^+$, then it is well known (see e.g. [58, Lemma 7.2.2]) that we can find $x_n \in L_+^2$, $n \in \mathbb{N}$, such that $x_n \to x$ in H^{-1}, hence by the last part of Theorem 2.5.1, $X(\cdot, x_n) \to X(\cdot, x)$ in $L_W^2(\Omega, C([0, T]; H^{-1}))$ and the assertion follows. $\qquad\square$

2.7 Comments and Bibliographical Remarks

In an appropriate form, Theorem 2.5.1 was first established for the two-phase Stefan problem with additive noise in [10]. It should be said, however, that Eq. (2.1) with Lipschitz monotone β is relevant as well in the description of a more general class of phase-transition models perturbed by Gaussian noise.

The method to get existence was to approximate the equation by one with Lipschitz nonlinearities F_ϵ in the basic space H^{-1} via Yosida approximations for nonlinear maximal monotone operators. It should be said that this approximation technique, though largely used, is not the only possible in nonlinear infinite dimensional analysis. Another one is the finite dimensional approximation via the Faedo–Galerkin method which leads also in this case to comparable results. However, there are some obvious advantage of the Yosida approximations and the most important one is that it does not change the basic functional space.

We also note that the results of this chapter extend mutatis mutandis to time dependent stochastic equations

$$dX = \Delta(\beta(t, X(t))dt + \sigma(X(t))dW(t),$$

where $\beta : [0, T] \times \mathbb{R} \to \mathbb{R}$ is measurable and for fixed t monotonically increasing and Lipschitz in $r \in \mathbb{R}$ uniformly in $t \in [0, T]$.

Appendix: Two Analytical Inequalities

Let us consider the Laplace operator in $L^2(\mathcal{O})$, $\mathcal{O} \in \mathbb{R}^d$, with homogeneous boundary conditions and its orthonormal basis of eigenfunctions, that is

$$- \Delta e_k = \alpha_k e_k \text{ in } \mathcal{O}, \quad e_k = 0 \text{ on } \partial\mathcal{O}. \tag{2.82}$$

We set $f_k = \alpha_k^{1/2} e_k$ so that $\{f_k\}$ is an orthonormal basis in $H^{-1}(\mathcal{O})$. We assume that $\partial\mathcal{O}$ is sufficiently regular (for instance of class C_2) in order to apply [66].

Proposition 1 *There exist $C_1 > 0$ and $C_2 > 0$ such that*

$$\|xf_k\|_{-1} \leq C_1 \alpha_k^d \|x\|_{-1}^2, \quad \forall \, k \in \mathbb{N}. \tag{2.83}$$

and

$$|xe_k|_2^2 \leq C_2 \alpha_k^{d-1} |x|_2^2, \quad \forall \, k \in \mathbb{N}, \tag{2.84}$$

Proof The proof of (2.84) is very simple. In fact for each $x \in L^2$ we have

$$|xe_k|_2 \leq |x|_2 |e_k|_\infty \leq c\alpha_k^{\frac{d-1}{2}} |x|_2, \quad \forall \, k \in \mathbb{N},$$

because by [66] we have $|e_k|_\infty \leq c\alpha_k^{\frac{d-1}{2}}$ for all $k \in \mathbb{N}$.

Let us now consider (2.83). Since H^{-1} is the dual of H_0^1 we have

$$|xe_k|_{-1}^2 = \sup \left\{ |\langle xe_k, \varphi \rangle|_2^2 : \varphi \in H_0^1, \, |\varphi|_{H_0^1} \leq 1 \right\}. \tag{2.85}$$

But

$$|\langle xe_k, \varphi \rangle|_2^2 = |\langle x, e_k\varphi \rangle|_2^2 \leq |x|_{-1}^2 |e_k\varphi|_{H_0^1}^2$$

On the other hand, for all $k \in \mathbb{N}$

$$|e_k\varphi|_{H_0^1}^2 = |\nabla(e_k\varphi)|_2^2 = - \int_{\mathcal{O}} e_k \varphi \, \Delta(e_k \varphi) \, d\xi$$

$$= - \int_{\mathcal{O}} (e_k \varphi^2 \, \Delta e_k + e_k^2\varphi \, \Delta\varphi + \tfrac{1}{2}\nabla(e_k^2) \cdot \nabla(\varphi^2)) d\xi$$

$$= - \int_{\mathcal{O}} (e_k \varphi^2 \, \Delta e_k + e_k^2\varphi \, \Delta\varphi - \tfrac{1}{2} e_k^2 \, \Delta(\varphi^2)) d\xi$$

Since

$$\Delta(\varphi^2) = 2\varphi \, \Delta\varphi + 2|\nabla\varphi|^2,$$

we have

$$|e_k\varphi|^2_{H^1_0} = \int_{\mathscr{O}}(\alpha_k\varphi^2 + |\nabla\varphi|^2))e_k^2 d\xi, \quad \forall\, k \in \mathbb{N}. \tag{2.86}$$

Therefore,

$$|e_k\varphi|^2_{H^1_0} \le \alpha_k|\varphi e_k|^2_2 + |\varphi|^2_{H^1_0}|e_k|^2_\infty, \quad \forall\, k \in \mathbb{N}. \tag{2.87}$$

Now by the Sobolev embedding theorem we have $H^1_0 \subset L^{\frac{2d}{d-2}}$ for $d > 3$, $H^1_0 \subset \cap_{p\ge2}L^p$ for $d = 1,2$ with continuous embedding. Then, using Hölder in the first term of (2.87) we see that there is a constant $c > 0$ such that

$$|e_k\varphi|^2_{H^1_0} \le (c\alpha_k|e_k|^2_d + |e_k|^2_\infty)|\varphi|^2_{H^1_0(\mathscr{O})}. \tag{2.88}$$

Now as mentioned earlier we know that

$$|e_k|^2_\infty \le c_1\alpha_k^{d-1}, \quad \forall\, k \in \mathbb{N},$$

so that, by interpolation[3]

$$|e_k|^2_d \le c_2\alpha_k^{\frac{(d-1)(d-2)}{d}}, \quad \forall\, k \in \mathbb{N},$$

Finally, we find

$$|e_k\varphi|^2_{H^1_0} \le c(\alpha_k^{1+\frac{(d-1)(d-2)}{d}} + \alpha_k^{d-1})|\varphi|^2_{H^1_0} \le c_1\alpha_k^{d-1}|\varphi|^2_{H^1_0}, \quad \forall\, k \in \mathbb{N}, \tag{2.89}$$

and therefore by (2.85) (2.83) follows.

---

[3] $|f|_p \le |f|_2^{\frac{2}{p}}\,|f|_\infty^{\frac{p-2}{p}}.$

Chapter 3
Equations with Maximal Monotone Nonlinearities

We shall study here Eq. (1.1) for general (multivalued) maximal monotone graphs $\beta : \mathbb{R} \to 2^{\mathbb{R}}$ with polynomial growth. The principal motivation for the study of these equations comes from nonlinear diffusion models presented in Sect. 1.1.

3.1 Introduction and Setting of the Problem

We are here concerned with the equation

$$\begin{cases} dX(t) = \Delta\beta(X(t))dt + \sigma(X(t))dW(t), \\ x(0) = x \in H^{-1}, \end{cases} \tag{3.1}$$

where β is a maximal monotone graph in $\mathbb{R} \times \mathbb{R}$ (possibly multivalued).

More precisely we assume that

Hypothesis 4 *(i) β is a maximal monotone graph on $\mathbb{R} \times \mathbb{R}$ such that $0 \in \beta(0)$. There exist $C > 0$ and $m \geq 0$ such that*

$$\sup\{|\theta| : \ \theta \in \beta(r)\} \leq C(1 + |r|^m), \quad \forall \, r \in \mathbb{R}. \tag{3.2}$$

(ii) σ is Lipschitzian from H^{-1} into $\mathscr{L}_2(H^{-1})$, so that (2.6) (2.5) hold.
(iii) σ is Lipschitzian from L^2 into $\mathscr{L}_2(H^{-1}, L^2)$, so that (2.8) (2.7) hold.
(iv) There exists $\sigma_3 > 0$ and $K_3 > 0$ such that

$$\sum_{k=1}^{\infty}[(\sigma(x)f_k)(\xi)]^2 \leq \sigma_3^2 + K_3^2|x(\xi)|^2, \quad \forall \, x \in L^2, \xi \in \mathcal{O}, \tag{3.3}$$

© Springer International Publishing Switzerland 2016
V. Barbu et al., *Stochastic Porous Media Equations*, Lecture Notes
in Mathematics 2163, DOI 10.1007/978-3-319-41069-2_3

where $\{f_k\}$ is an orthonormal basis on H^{-1} of eigenfunctions of A.
(v) W is a cylindrical Wiener process in H^{-1} of the form (2.10).

Remark 3.1.1 Hypothesis 4(ii) and (iii) coincide with Hypothesis 1(ii) and (iii) respectively. They have been discussed in Examples 2.1.1 and 2.1.2. Hypothesis 4(iv) is just Hypothesis 2 which has been discussed in Example 2.4.5.

Let us recall that when $\sigma(x) = A^{-\gamma/2}$, Hypothesis 4(ii) and (iii) are fulfilled provided $\gamma > \frac{d+2}{2}$ whereas Hypothesis 4(iv) is fulfilled provided

$$\gamma > \frac{3d}{2}. \tag{3.4}$$

Finally, in the case of linear noise of type $\sigma(x) = kxA^{-\gamma/2}$ Hypothesis 4(ii)–(iv) are fulfilled provided $\gamma > \frac{3d}{2}$.   □

In this chapter we shall prove existence and uniqueness of a solution X of Eq. (3.1) in a sense to be made precise in Definition 3.1.2 below. Under Hypothesis 4 we are not able to show in general existence of a strong solution X that is such that $\beta(X) \in H_0^1$. However, this is possible in the important case $\beta(r) = r^{2m+1}$ of *slow diffusions*. This case will be treated in Sect. 3.6. In Sect. 3.7 we shall consider the *fast diffusions* case of $\beta(r) = r^\alpha$ with $\alpha \in [0, 1)$. Here, under a suitable further assumption, we show that solutions of (3.1) has a non zero probability of extinction.

Definition 3.1.2 Let $x \in H^{-1}$. By a *distributional solution* to (3.1) on $[0, T]$ we mean a stochastic process X such that

(i) $X \in L_W^2(\Omega; C([0, T]; H^{-1})) \cap L^{m+1}(\Omega \times (0, T) \times \mathcal{O})$.
(ii) There exists a process $Z : [0, T] \to H^{-1}$ such that $Z \in L^{\frac{m+1}{m}}(\Omega \times (0, T) \times \mathcal{O})$, $Z \in \beta(X)$ a.e. in $\Omega \times (0, T) \times \mathcal{O}, \int_0^t Z(s)ds \in C([0, T]; H_0^1)$ and $\mathbb{P}$-a.s.

$$\langle X(t), f_j\rangle_{-1} = \langle x, f_j\rangle_{-1} - \int_0^t {}_{\frac{m+1}{m}}\langle Z(s), f_j\rangle_{m+1} \, ds$$

$$+ \sum_{k=1}^\infty \int_0^t \langle \sigma(X(s))f_k, f_j\rangle_{-1} \, dW_k(s), \quad \forall j \in \mathbb{N}, \ t \in [0, T],$$

$$\tag{3.5}$$

where $\{f_k\}$ is an eigenbasis for $A = -\Delta$ in H^{-1} with corresponding eigenvalues $\{\alpha_k\}$.

We note that since f_j is regular, the term ${}_{\frac{m+1}{m}}\langle Z(s), f_j\rangle_{m+1}$ is well defined. We also note in particular that the distributional solution is $\mathbb{P}$-a.s., H^{-1}-valued continuous on $[0, T]$.

We call such a solution *distributional* because, though from the stochastic point of view the solution X given by Definition 3.1.2 is a strong one, from the PDE point of view it is a solution in the sense of distributions since the boundary condition $\beta(X) \ni 0$ on $\partial \mathcal{O}$ is satisfied in a weak sense only. (See, however, Remark 3.1.4.)

□

Remark 3.1.3 We note that the self-organized criticality equation (1.8) is covered by Hypothesis 4. Indeed in this case we have

$$\beta(r) = \text{sign } r + \nu(r),$$

where ν is a maximal monotone graph which satisfies the growth condition (3.2). (In most notable situations $\nu(r) = ar$ where $a \geq 0$.)   □

Remark 3.1.4 Equation (3.5) can be equivalently written as

$$X(t) = x - \Delta \int_0^t Z(s)ds + \int_0^t \sigma(X(s))dW(s), \quad t \in [0, T], \tag{3.6}$$

where $\Delta : H_0^1 \to H^{-1}$ is taken in the sense of distributions on $\mathcal{O}$. In this form the difference between distributional and strong solutions to (3.1) is apparent.   □

3.2 Uniqueness

We show first Itô's formula for the squared H^{-1} norm of a solution of (3.1).

Proposition 3.2.1 *Let* $X \in L_W^2(\Omega; C([0,T]; H^{-1})) \cap L^{m+1}(\Omega \times (0,T) \times \mathcal{O})$ *be a distributional solution of* (3.1) *and let* Z *as in definition 3.1.2(ii). Then we have*

$$\mathbb{E}\|X(t)\|_{-1}^2 + 2\mathbb{E}\int_0^t \tfrac{m+1}{m} \langle Z(s), X(s)\rangle_{m+1}\, ds$$

$$\tag{3.7}$$

$$= \|x\|_{-1}^2 + \mathbb{E}\int_0^t \|\sigma(X(s))\|_{\mathscr{L}_2(H^{-1})}^2\, ds$$

Proof For any $j \in \mathbb{N}$, we see from (3.5) that $\langle X(\cdot), f_j\rangle_{-1}$ is an Itô's process and that

$$d\langle X(\cdot), f_j\rangle_{-1} = -\tfrac{m+1}{m} \langle Z(t), f_j\rangle_{m+1}\, dt + \sum_{k=1}^{\infty} \langle \sigma(X(t))f_k, f_j\rangle_{-1}\, dW_k(t).$$

It follows that

$$d|\langle X(t),f_j\rangle_{-1}|^2 = -2\langle X(t),f_j\rangle_{-1}\,\,{}_{\frac{m+1}{m}}\langle Z(t),f_j\rangle_{m+1}\,dt$$

$$+2\sum_{k=1}^{\infty}\langle X(t),f_j\rangle_{-1}\langle\sigma(X(t))f_k,f_j\rangle_{-1}\,dW_k(t)) \qquad (3.8)$$

$$+\sum_{k=1}^{\infty}|\langle\sigma(X(t))f_k,f_j\rangle_{-1}|^2\,dt.$$

We note that since $f_j = \sqrt{\alpha_j}\,e_j$, we have

$$\langle X(t),f_j\rangle_{-1}\,\,{}_{\frac{m+1}{m}}\langle Z(t),f_j\rangle_{m+1} = \int_{\mathscr{O}} X(t)e_j\,d\xi\int_{\mathscr{O}}Z(t)e_j\,d\xi.$$

Hence integrating (3.8) with respect to t and taking expectation, it follows that

$$\mathbb{E}|\langle X(t),f_j\rangle_{-1}|^2 = |\langle x,f_j\rangle_{-1}|^2 - 2\mathbb{E}\int_{\mathscr{O}}X(t)f_j\,d\xi\int_{\mathscr{O}}Z(t)f_j\,d\xi$$

$$+\sum_{k=1}^{\infty}\mathbb{E}\int_0^t|\langle\sigma(X(t))f_k,f_j\rangle_{-1}|^2\,ds.$$

Now the conclusion follows summing up in j because $X \in L^{m+1}(\Omega\times(0,T)\times\mathscr{O})$ and $Z \in L^{\frac{m+1}{m}}(\Omega\times(0,T)\times\mathscr{O})$.

The last conclusion is, of course, only rigorous if $m = 1$. But applying $(1+\epsilon A)^{-l}$ to (3.6), arguing as above with $(1+\epsilon A)^{-l}X(t)$, l large enough, replacing $X(t)$ and letting eventually ϵ tend to zero, one obtains the assertion. $\qquad\square$

Theorem 3.2.2 *Equation* (3.1) *has at most one distributional solution.*

Proof Let X_1 and X_2 be two distributional solutions to Eq. (3.1) and let Z_1 and Z_2 be as in Definition 3.1.2(ii). Then, arguing as in the proof of Proposition 3.2.1 we find

$$\mathbb{E}\|X(t) - Y(t)\|_{-1}^2 + 2\mathbb{E}\int_0^t {}_{\frac{m+1}{m}}\langle(Z(s)-Z_1(s),X(s)-X_1(s)\rangle_{m+1}ds$$

$$= \mathbb{E}\int_0^t\|\sigma(X(s))-\sigma(X_1(s))\|_{\mathscr{L}_2(H^{-1})}^2\,ds$$

Since β is (maximal) monotone, it follows that

$$\mathbb{E}\|X(t) - Y(t)\|_{-1}^2 \le \mathbb{E}\int_0^t\|\sigma(X(s))-\sigma(X_1(s))\|_{\mathscr{L}_2(H^{-1})}^2\,ds. \qquad (3.9)$$

Now, by Hypothesis 4(iii) we obtain

$$\mathbb{E}\|X(t) - Y(t)\|_{-1}^2 \le K_1 \mathbb{E} \int_0^t \|X(s) - Y(s)\|_{-1}^2 \, ds, \quad \forall \, t \in [0, T],$$

which by Gronwall's lemma yields $X = Y$.   □

3.3 The Approximating Problem

For any $\epsilon > 0$ we consider the equation

$$\begin{cases} dX_\epsilon(t) = \Delta(\beta_\epsilon(X_\epsilon(t)) + \epsilon X_\epsilon(t))dt + \sigma(X_\epsilon(t)) \, dW(t), \\ X_\epsilon(0) = x, \end{cases} \tag{3.10}$$

where β_ϵ is the Yosida approximation of β, that is

$$\beta_\epsilon(r) = \frac{1}{\epsilon} (r - J_\epsilon(r)) \in \beta(J_\epsilon(r)) \tag{3.11}$$

and

$$J_\epsilon(r) = (1 + \epsilon\beta)^{-1}(r), \quad r \in \mathbb{R}. \tag{3.12}$$

We set in the following

$$\tilde{\beta}_\epsilon(r) = \beta_\epsilon(r) + \epsilon r. \tag{3.13}$$

Since β_ϵ is Lipschitz and nondecreasing, by Theorem 2.5.1 problem (3.10) has a unique generalized solution $X_\epsilon \in L_W^2(\Omega; C([0, T]; H^{-1}))$, which in addition is strong (in the sense of Definition 2.1.4) if $x \in L^2$ (that is $(\beta_\epsilon + \epsilon)(X_\epsilon I(t))$ belongs to $L_W^2(\Omega; L^2(0, T; H_0^1))$.) Moreover, this implies that $X_\epsilon \in L_W^2(\Omega; L^2(0, T; H_0^1))$

Remark 3.3.1 It should be noticed that $G_\epsilon := \Delta\beta_\epsilon$, $\epsilon > 0$, *is not* the Yosida approximations of the maximal monotone operator $F = -\Delta\beta$ in H^{-1}.

We need the following simple lemma.

Lemma 3.3.2 *We have*

$$|\beta_\epsilon(r)| \le C(|r|^m + 1), \quad \forall \, r \in \mathbb{R}, \, \epsilon > 0. \tag{3.14}$$

Proof We have in fact

$$|\beta_\epsilon(r)| \le \sup\{|\theta| : \theta \in \beta(J_\epsilon(r))\} \le C(|J_\epsilon(r)|^m + 1) \le C(|r|^m + 1). \quad □$$

3.3.1 Estimating $\mathbb{E}\|X_\epsilon(t)\|^2_{-1}$

This estimate is a standard consequence of the monotonicity of $F = -\Delta\beta$. We present here a proof for the readers convenience, though it is very similar to that of Lemma 2.4.2.

Proposition 3.3.3 *Assume that Hypothesis 4(i),(ii),(iii),(v) are fulfilled. Then for each $x \in L^2$ we have*

$$\mathbb{E}\|X_\epsilon(t)\|^2_{-1} + 2\mathbb{E}\int_0^t \langle X_\epsilon(s), \tilde{\beta}_\epsilon(X_\epsilon(s))\rangle_2\, ds \tag{3.15}$$

$$\leq e^{K_1^2 t}(\|x\|^2_{-1} + 2\sigma_1^2 t)$$

and[1]

$$\mathbb{E} \sup_{t\in[0,T]} \|X_\epsilon(t)\|^2_{-1} \leq 2(\|x\|^2_{-1} + 2\sigma_1^2 T) \tag{3.16}$$

$$+ (2M_1 + 16c_1^2 M_1^2)\mathbb{E}\int_0^T (1 + \|X_\epsilon(s)\|^2_{-1})ds.$$

If $x \in H^{-1}$, then (3.16) still holds and likewise (3.15), but without the integral term on the left hand side (which is not well defined since $X_\epsilon(s) \notin L^2$).

Proof We only consider the case $x \in L^2$. If merely $x \in H^{-1}$, the assertion follows by approximation. By Itô's formula applied to $\|X_\epsilon(t)\|^2_{-1}$, we have

$$d\|X_\epsilon(t)\|^2_{-1} + 2\langle X_\epsilon(t), \tilde{\beta}_\epsilon(X_\epsilon(t))\rangle_2\, dt = \|\sigma(X_\epsilon(t))\|^2_{\mathscr{L}_2(H^{-1})}\, dt \tag{3.17}$$

$$+ 2\langle X_\epsilon(t), \sigma(X_\epsilon(t))dW(t)\rangle_{-1}.$$

Consequently, integrating with respect to t, yields

$$\|X_\epsilon(t)\|^2_{-1} + 2\int_0^t \langle X_\epsilon(s), \tilde{\beta}_\epsilon(X_\epsilon(s))\rangle_2\, ds$$

$$= \|x\|^2_{-1} + \int_0^t \|\sigma(X_\epsilon(s))\|^2_{\mathscr{L}_2(H^{-1})}\, ds + 2\int_0^t \langle X_\epsilon(s), \sigma(X_\epsilon(s))dW(s)\rangle_{-1}. \tag{3.18}$$

---

[1] c_1 is the constant from the Burkholder–Davis–Gundy inequality (1.23).

Now taking expectation, yields

$$\mathbb{E}\|X_\epsilon(t)\|_{-1}^2 + 2\mathbb{E}\int_0^t \langle X_\epsilon(s), \tilde{\beta}_\epsilon(X_\epsilon(s))\rangle_2 ds$$

$$= \|x\|_{-1}^2 + \mathbb{E}\int_0^t \|\sigma(X_\epsilon(s))\|_{\mathscr{L}_2(H^{-1})}^2 ds. \tag{3.19}$$

Finally, differentiating with respect to t and recalling Hypothesis 4(iii), we find

$$\frac{d}{dt}\,\mathbb{E}\|X_\epsilon(t)\|_{-1}^2 + 2\mathbb{E}\langle X_\epsilon(t), \tilde{\beta}_\epsilon(X_\epsilon(t))\rangle_2$$

$$= \mathbb{E}\|\sigma(X_\epsilon(t))\|_{\mathscr{L}_2(H^{-1})}^2 \le 2\sigma_1^2 + 2K_1^2\mathbb{E}\|X_\epsilon(t)\|_{-1}^2.$$

Then (3.15) follows.

Let us prove (3.16). By (3.18) it follows that

$$\sup_{t\in[0,T]} \|X_\epsilon(t)\|_{-1}^2 \le \|x\|_{-1}^2 + 2\sigma_1^2 + 2K_1^2\mathbb{E}\int_0^T \|X_\epsilon(s)\|_{-1}^2)ds$$

$$+2\sup_{t\in[0,T]}\left|\int_0^t \langle X_\epsilon(s)\sigma(X_\epsilon(s)), dW(s)\rangle_{-1}\right|. \tag{3.20}$$

Arguing as in the proof of Lemma 2.4.2 (see (2.44) and the following), yields

$$\mathbb{E}\sup_{t\in[0,T]} \|X_\epsilon(t)\|_{-1}^2 \le \|x\|_{-1}^2 + \mathbb{E}\int_0^T (\sigma_1^2 + K_1^2\|X_\epsilon(s)\|_{-1}^2)ds$$

$$+\frac{1}{2}\,\mathbb{E}\sup_{t\in[0,T]} \|X_\epsilon(t)\|_{-1}^2 + 8c_1^2\mathbb{E}\int_0^T (\sigma_1^2 + K_1^2\|X_\epsilon(s)\|_{-1}^2)ds.$$

It follows that

$$\mathbb{E}\sup_{t\in[0,T]} \|X_\epsilon(t)\|_{-1}^2 \le 2\|x\|_{-1}^2 + 2\mathbb{E}\int_0^T (\sigma_1^2 + K_1^2\|X_\epsilon(s)\|_{-1}^2)ds$$

$$+16c_1^2\,\mathbb{E}\int_0^T (\sigma_1^2 + K_1^2\|X_\epsilon(s)\|_{-1}^2)ds, \tag{3.21}$$

which yields (3.16). $\qquad\square$

3.3.2   Estimating $\mathbb{E}|X_\epsilon(t)|_p^p$

Here we need the new Hypothesis 3.1(iv).

Proposition 3.3.4 *Assume that Hypothesis 3.1 is fulfilled and let $x \in L^p$ with $p \geq 2$. Then there exists a constant $M_{1,p} > 0$ such that*

$$\sup_{t \in [0,T]} \mathbb{E}|X_\epsilon(t)|_p^p \leq e^{M_{1,p}t}(|x|_p^p + M_{1,p}T). \tag{3.22}$$

Proof We just apply (2.60) to the strong solution X_ϵ to problem (3.10).   □

3.4   Solution to Problem (3.1)

Theorem 3.4.1 *Assume that Hypothesis 4 is fulfilled. Then for each $T > 0$ and all $x \in L^{2m} \cap L^2$, problem (3.1) has a unique distributional solution X. Moreover, if $x \in H^{-1}$ then problem (3.1) has a unique generalized solution. If in addition Hypothesis 3 is fulfilled and $x \in L_+^p \cap L^2$, $p \geq 2m$, we have $X(t)(\omega) \in L_+^p$ for $dt \otimes \mathbb{P}$-a.e. $(t, \omega) \in [0, T] \times \Omega$. Furthermore, if $x \in H^{-1} \cap \mathscr{M}^+$, then $\mathbb{P}$-a.s. we have for the generalized solution X that $X(t) \in \mathscr{M}^+ \cap H^{-1}$ for all $t \in [0, T]$.*

Here $L_+^p = \{x \in L^p : x \geq 0, \text{ a.e. in } \mathscr{O}\}$.

Proof Step 1. For each $x \in L^{2m}$ the sequence $\{X_\epsilon\}$ (where X_ϵ is the solution of (3.10)) is Cauchy in $C_W([0, T]; L^2(\Omega; H^{-1}))$.

Indeed if $\epsilon, \eta > 0$ then by Itô's formula we have[2]

$$d\|X_\epsilon(t) - X_\eta(t)\|_{-1}^2 + 2\langle X_\epsilon(t) - X_\eta(t), \widetilde{\beta}_\epsilon(X_\epsilon(t)) - \widetilde{\beta}_\eta(X_\eta(t))\rangle_2 dt$$

$$= 2\langle X_\epsilon(t) - X_\eta(t), (\sigma(X_\epsilon(t)) - \sigma(X_\eta(t))dW(t)\rangle_{-1} \tag{3.23}$$

$$+ \|\sigma(X_\epsilon(t)) - \sigma(X_\eta(t))\|_{\mathscr{L}_2(H^{-1})}^2 dt.$$

We use now the identity

$$(\beta_\epsilon(r) - \beta_\eta(s))(r - s) = (\beta(J_\epsilon(r)) - \beta(J_\eta(s))(r - s)$$

$$= (\beta(J_\epsilon(r)) - \beta(J_\eta(s))(J_\epsilon(r) - J_\epsilon(s)) \tag{3.24}$$

$$+ (\beta_\epsilon(r) - \beta_\eta(s))(\epsilon\beta_\epsilon(r) - \eta\beta_\eta(s)),$$

---

[2]Recall that $\widetilde{\beta}_\epsilon(r) = \beta_\epsilon(r) + \epsilon r$ and $\widetilde{\beta}_\eta(r) = \beta_\eta(r) + \eta r, r \in \mathbb{R}$.

to obtain that

$$d\|X_\epsilon(t) - X_\eta(t)\|_{-1}^2 + 2\langle J_\epsilon(X_\epsilon(t)) - J_\eta(X_\eta(t)), \beta(J_\epsilon(X_\epsilon(t))) - \beta(J_\eta(X_\eta(t))))\rangle_2 dt$$

$$+ 2\langle \epsilon X_\epsilon(t) - \eta X_\eta(t), X_\epsilon(t) - X_\eta(t)\rangle_2$$

$$+ 2\langle (\beta_\epsilon(X_\epsilon(t)) - \beta_\eta(X_\eta(t)))(\epsilon\beta_\epsilon(X_\epsilon(t)) - \eta\beta_\eta(X_\eta(t)))\rangle_2$$

$$+ 2\langle X_\epsilon(t) - X_\eta(t), (\sigma(X_\epsilon(t)) - \sigma(X_\eta(t))dW(t)\rangle_{-1}$$

$$= \|\sigma(X_\epsilon(t)) - \sigma(X_\eta(t)\|_{\mathscr{L}_2(H^{-1})}^2 dt.$$

$$(3.25)$$

Integrating with respect to t and taking expectation, yields

$$\mathbb{E}\|X_\epsilon(t) - X_\eta(t)\|_{-1}^2$$

$$+ \mathbb{E}\int_0^t \langle J_\epsilon(X_\epsilon(s)) - J_\eta(X_\eta(s)), \beta(J_\epsilon(X_\epsilon(s))) - \beta(J_\eta(X_\eta(s)))\rangle_2 \, ds$$

$$+ \mathbb{E}\int_0^t \langle \beta_\epsilon(X_\epsilon(s)) - \beta_\eta(X_\eta(s)))(\epsilon\beta_\epsilon(X_\epsilon(s)) - \eta\beta_\eta(X_\eta(s)))\rangle_2 \, ds$$

$$+ \mathbb{E}\int_0^t \langle \epsilon X_\epsilon(s) - \eta X_\eta(s), X_\epsilon(s) - X_\eta(s)\rangle_2 ds$$

$$= \mathbb{E}\int_0^t \|\sigma(X_\epsilon(s)) - \sigma(X_\eta(s)\|_{\mathscr{L}_2(H^{-1})}^2 ds.$$

$$(3.26)$$

To go further we need an estimate for

$$\mathbb{E}\int_{\mathscr{O}} (\beta_\epsilon(X_\epsilon(s)))^2 d\xi$$

Recalling Lemma 3.3.2 it is enough estimating

$$\mathbb{E}\int_{\mathscr{O}} (X_\epsilon(s))^{2m} d\xi = \mathbb{E}|X_\epsilon(s)|_{2m}^{2m}$$

Now, thanks to Proposition 3.3.4 for $p = 2m$, we have

$$\mathbb{E}|X_\epsilon(t)|_{2m}^{2m} \le e^{M_{1,2m}t}(|x|_{2m}^{2m} + M_{1,2m}T).$$

$$(3.27)$$

so that by (2.6) we have

$$\mathbb{E}\|X_\epsilon(t) - X_\eta(t)\|_{-1}^2$$

$$+\mathbb{E}\int_0^t \langle J_\epsilon(X_\epsilon(s)) - J_\eta(X_\eta(s)), \beta(J_\epsilon(X_\epsilon(s))) - \beta(J_\eta(X_\eta(s)))\rangle_2 ds$$

$$\leq 4(\epsilon + \eta)e^{M_{2m}T}\left(|x|_{2m}^{2m} + M_{2m}\right) + K_1\mathbb{E}\int_0^t \|X_\epsilon(s) - X_\eta(s)\|_{-1}^2 ds$$

$$+\delta(\epsilon,\eta).$$

It follows that

$$\mathbb{E}\|X_\epsilon(t) - X_\eta(t)\|_{-1}^2$$

$$+\mathbb{E}\int_0^t e^{K_1(t-s)}\langle J_\epsilon(X_\epsilon(s)) - J_\eta(X_\eta(s)), \beta(J_\epsilon(X_\epsilon(s))) - \beta(J_\eta(X_\eta(s)))\rangle_2 ds$$

$$\leq 4(\epsilon + \eta)e^{M_{2m}T + K_1 T}\left(|x|_{2m}^{2m} + M_{2m}\right). \tag{3.28}$$

Therefore $\{X_\epsilon\}$ is Cauchy in $C_W([0,T]; L^2(\Omega; H^{-1}))$ as claimed.

Step 2. For each $x \in L^{2m}$, $\{X_\epsilon\}$ is Cauchy in $L_W^2(\Omega; C([0,T]; (H^{-1}))$.

It is similar to the proof of Step 2 of Theorem 2.5.1. Namely, one applies to (3.25) the Burkholder–Davis–Gundy inequality and obtains the desired conclusion.

We define now

$$X = \lim_{\epsilon \to 0} X_\epsilon \quad \text{in } L_W^2(\Omega; C([0,T]; H^{-1})), \tag{3.29}$$

which exists by (3.28).

Step 3. For $x \in L^{2m}$, X is a distributional solution to Eq. (3.1).

By inequality (3.22) we have

$$\mathbb{E}|X_\epsilon(t)|_{m+1}^{m+1} \leq e^{M_{m+1}t}\left(|x|_{m+1}^{m+1} + M_{m+1}T\right). \tag{3.30}$$

From (3.30) it follows that for a subsequence $\{\epsilon\} \to 0$ we have

$$\begin{aligned} X_\epsilon &\to X \quad \text{weakly in } L^{m+1}(\Omega \times (0,T) \times \mathcal{O}), \\ X_\epsilon &\to X \text{ weak* in } L^\infty(0,T; L^{m+1}(\Omega \times \mathcal{O}))). \end{aligned} \tag{3.31}$$

Moreover from by (3.15) it follows that for a subsequence $\{\epsilon\} \to 0$

$$\beta_\epsilon(X_\epsilon) \to Z \quad \text{weakly in } L^{\frac{m+1}{m}}(\Omega \times (0,T) \times \mathcal{O}). \tag{3.32}$$

If we rewrite (3.10) as

$$(-\Delta)^{-1}X_\epsilon(t) = (-\Delta)^{-1}x + \int_0^t \tilde{\beta}_\epsilon(X_\epsilon(s))ds + (-\Delta)^{-1}\int_0^t \sigma(X_\epsilon(s))dW(s)$$

and take into account (3.29) we get

$$\int_0^t Z(s)ds = (-\Delta)^{-1}x - (-\Delta)^{-1}\int_0^t \sigma(X_\epsilon(s))dW(s) \in H_0^1, \quad \forall t \in [0,T].$$

On the other hand, we have

$$d(X_\lambda(t) - X_\mu(t)) - \Delta(\beta_\lambda(X_\lambda(t)) - \beta_\mu(X_\mu(t)) + \lambda X_\lambda(t) - \mu X_\mu(t))dt$$
$$= (\sigma(X_\lambda(t)) - \sigma(X_\mu(t)))dW(t)$$

In order to complete the proof of existence, it suffices to show that

$$Z(\omega,t,\xi) \in \beta(X(\omega,t,\xi)) \quad \text{a.e in } \Omega \times (0,T) \times \mathcal{O}. \tag{3.33}$$

Since the operator

$$L^{m+1}(\Omega \times (0,T) \times \mathcal{O}) \to L^{\frac{m+1}{m}}(\Omega \times (0,T) \times \mathcal{O}), \quad X \to \beta(X),$$

is maximal monotone, in the duality pair

$$\left(L^{m+1}(\Omega \times (0,T) \times \mathcal{O}), L^{m+1}(\Omega \times (0,T) \times \mathcal{O})' = L^{\frac{m+1}{m}}(\Omega \times (0,T) \times \mathcal{O})\right),$$

it suffices to show that (see Proposition 1.2.8)

$$\liminf_{\epsilon \to 0} \mathbb{E}\int_0^T \int_\mathcal{O} \beta_\epsilon(X_\epsilon)X_\epsilon d\xi dt \leq \mathbb{E}\int_0^T \int_\mathcal{O} ZX d\xi dt. \tag{3.34}$$

To prove (3.34) we note that letting $\epsilon \to 0$ in (3.19) we have

$$2\liminf_{\lambda \to 0} \mathbb{E}\int_0^t \langle X_\epsilon(s), \beta_\epsilon(X_\epsilon(s))\rangle_2 ds$$

$$\tag{3.35}$$

$$= \|x\|_{-1}^2 + \int_0^t \|\sigma(X(s))\|_{\mathscr{L}_2(H^{-1})}^2 ds - \mathbb{E}\|X(t)\|_{-1}^2.$$

so that the conclusion follows from (3.7).

By Itô's formula it follows for $x, y \in L^{2m}$

$$\mathbb{E}\sup_{t \in [0,T]} \|X(t,x) - X(t,y)\|_{-1}^2 \leq C\|x - y\|_{-1}^2,$$

where $X = X(t, x)$ is the solution to (3.1). This implies by density existence for $x \in H^{-1}$.

The last part of the assertion follows by (3.31), (3.32) and by the same arguments as at the end of the proof of Theorem 2.6.2.   □

Remark 3.4.2 If in the situation of Theorem 3.4.1, β is single-valued and $r\beta(r) \geq cr^2$, there exists a unique distributional (not just generalized) solution for all $x \in H^{-1}$. This follows as a special case of [84, Theorem 3.9].   □

Remark 3.4.3 In particular, Theorem 3.4.1 applies for $\beta(r) = \rho\, \text{sign}\, r$, that is for the self-organized criticality equation (1.8).   □

3.5   Slow Diffusions

We are here concerned with Eq. (3.1) under the following hypothesis

Hypothesis 5 *(i)* $\beta : \mathbb{R} \to \mathbb{R}$ *is maximal monotone and there is* $m > 0$, $a > 0$
   such that $\beta(r)r \geq ar^{2m+2}$ *for all* $r \in \mathbb{R}$.
(ii) Hypothesis 4 is fulfilled.

As seen earlier in introduction, (3.1) is in this case the classical equation describing the diffusion in a porous medium. With respect to more general porous media equations treated previously, this one has some specific properties and peculiarities. In particular, admits strong solutions.

Definition 3.5.1 For any $x \in L^{2m+2}$ a *strong solution* to (3.1) on $[0, T]$ is a stochastic process X which belongs to $L^2_W(\Omega; C([0, T]; H^{-1}))$, such that $\beta(X) \in L^2(0, T; L^2(\Omega; H^1_0))$, and $\mathbb{P}$-a.s.

$$X(t) = x + \int_0^t \Delta\beta(X(s))ds + \int_0^t \sigma(X(s))dW(s), \quad \forall\, t \in [0, T]. \tag{3.36}$$

For any $x \in H^{-1}$ a generalized solution to (3.1) in $[0, T]$ is a process X which belongs to $L^2_W(\Omega; C([0, T]; H^{-1}))$ such that there exists $\{x_n\} \subset L^{2m+2}$ convergent to x in H^{-1} and

$$\lim_{n \to \infty} X_n = X \quad \text{in } L^2_W(\Omega; C([0, T]; H^{-1})),$$

where X_n is the strong solution to (3.1) with x_n replacing x.

We note that the strong solution X to (3.1) satisfies in particular the boundary Dirichlet conditions

$$\beta(X(t, \omega)) = 0 \quad \text{on } \partial\mathcal{O}, \quad \text{for } dt \otimes \mathbb{P}\text{-a.e. } (t, \omega) \in [0, T] \times \Omega.$$

which is not the case with distributional or generalized solutions.

3.5.1 The Uniqueness

The uniqueness of strong or generalized solutions follows from Theorem 3.2.2 because a strong solution is also a distributional solution. We prefer, however, to present another simpler proof

Proposition 3.5.2 *Equation* (3.1) *has at most one strong or generalized solution.*

Proof Let first X and Y be two strong solutions of (3.1). Then for F as defined in (2.3) by Itô's formula we have

$$d\|X(t) - Y(t)\|_{-1}^2 = -2\langle X(t) - Y(t), F(X(t)) - F(Y(t))\rangle_{-1}\, dt$$

$$+2\langle X(t) - Y(t), (\sigma(X(t)) - \sigma(Y(t)))dW(t)\rangle_{-1} \qquad (3.37)$$

$$+\|\sigma(X(t)) - \sigma(Y(t))\|_{\mathscr{L}_2(H^{-1})}^2\, dt, \qquad \mathbb{P}\text{-a.s.}$$

Taking into account the monotonicity of F and (2.4) it follows that

$$\mathbb{E}\|X(t) - Y(t)\|_{-1}^2 \le \mathbb{E}\int_0^t \|(\sigma(X(s)) - \sigma(Y(s)))\|_{\mathscr{L}_2(H^{-1})}^2 ds$$

$$\le K_1\mathbb{E}\int_0^t \|X(s) - Y(s)\|_{-1}^2\, ds.$$

Now the conclusion follows from Gronwall's lemma. Uniqueness of the generalized solution is also immediate, by definition. $\qquad\qquad\qquad\qquad\qquad\qquad \Box$

We are now going to show existence of a strong solution. For simplicity we shall take β of the special form

$$\beta(r) = a|r|^{2m}r, \quad \forall\, r \in \mathbb{R},$$

the general case being similar. The crucial point in the existence proof is an estimate of $\mathbb{E}\int_0^t \int_{\mathscr{O}} |D\beta_\epsilon(X_\epsilon(s))|^2\, d\xi\, ds$ which is provided by Proposition 3.5.4 below. First we prove a lemma.

Lemma 3.5.3 *We have*

$$0 \le \beta'_\epsilon(r) \le (2m+1)|r|^{2m}, \quad \forall\, r \in \mathbb{R},\ \epsilon > 0. \qquad (3.38)$$

Proof Let us set $s(r) = J_\epsilon(r)$, so that

$$s + \epsilon\beta(s) = r.$$

Differentiating with respect to r, yields

$$s'(r) + \epsilon\beta'(s)s'(r) = 1.$$

So

$$s'(r) = \frac{1}{1 + \epsilon\beta'(s)}.$$

Moreover, we have by (3.11)

$$\beta_\epsilon'(r) = \beta'(s)s'(r) = \frac{\beta'(s)}{1 + \epsilon\beta'(s)} \leq \beta'(s) = (2m + 1)|r|^{2m},$$

as claimed.   □

Now we are ready to prove the estimate for the solution X_ϵ to (3.1).

Proposition 3.5.4 *For any $x \in L^{2m+2}$ there exists a constant M_4 such that*

$$\mathbb{E}\int_0^t \int_{\mathscr{O}} |\nabla\beta_\epsilon(X_\epsilon(s))|^2\, d\xi\, ds \leq \frac{e^{M_4 t}}{(2m + 1)^2(2m + 2)}\left(|x|_{2m+2}^{2m+2} + M_4 T\right). \quad (3.39)$$

Proof Setting $p = 2m + 2$ and applying (3.22) we see that

$$\mathbb{E}\int_0^t \int_{\mathscr{O}} \beta_\epsilon'(X_\epsilon(s))|X_\epsilon(s)|^{2m}\, |\nabla X_\epsilon(s)|^2\, d\xi\, ds$$
$$\qquad\qquad\qquad\qquad\qquad\qquad\qquad\qquad\qquad\qquad\qquad\qquad\qquad (3.40)$$
$$\leq \frac{e^{M_{3,m+1}t}}{(2m + 1)(2m + 2)}\left(|x|_{2m+2}^{2m+2} + M_{3,m+1}T\right).$$

(We recall that $X_\epsilon \in L^\infty(0, T; L^p(\Omega; L^p)) \cap L^2(\Omega; L^2(0, T; H_0^1))$ and so the latter makes sense). On the other hand, taking into account (3.38), yields

$$\mathbb{E}\int_0^t \int_{\mathscr{O}} |\nabla\beta_\epsilon(X_\epsilon(s))|^2\, d\xi\, ds = \mathbb{E}\int_0^t \int_{\mathscr{O}}(\beta_\epsilon'(X_\epsilon(s)))^2|\nabla X_\epsilon(s)|^2\, d\xi\, ds$$

$$\leq (2m + 1)\mathbb{E}\int_0^t \int_{\mathscr{O}} \beta_\epsilon'(X_\epsilon(s))|X_\epsilon(s)|^{2m}|\nabla X_\epsilon(s)|^2\, d\xi\, ds.$$

Now by (3.40) we deduce (3.39).   □

Theorem 3.5.5 *Assume that Hypothesis 5 is fulfilled. Then for each $T > 0$ and all $x \in L^{2m+2}$ problem (3.1) has a unique strong solution. Moreover, if $x \in H^{-1}$ then problem (3.1) has a unique generalized solution. Finally, if in addition Hypothesis 3 is fulfilled and $x \in L_+^{2m+2}$, we have $X(t, \omega) \in L_+^{2m+2}$ for $dt \otimes \mathbb{P}$-a.e. $(t, \omega) \in [0, T] \times \Omega$. Furthermore, if $x \in H^{-1} \cap \mathscr{M}^+$, then $\mathbb{P}$-a.s. we have for the generalized solution X that $X(t) \in \mathscr{M}^+ \cap H^{-1}$ for all $t \in [0, T]$.*

Proof We already know by the proof of Theorem 3.4.1 that for each $x \in L^{2m+2}$ the sequence $\{X_\epsilon\}$ (where X_ϵ is the solution of (3.10)) is Cauchy in $L^2_W(\Omega; C([0, T]; (H^{-1}))$. Let $X \in L^2_W(\Omega; C([0, T]; H^{-1}))$ be such that

$$\lim_{\epsilon \to 0} X_\epsilon = X \quad \text{in } L^2_W(\Omega; C([0, T]; H^{-1})). \tag{3.41}$$

It remains to show that $\beta(X) \in L^2(0, T; L^2(\Omega; H^1_0)$ and that (3.36) is fulfilled. We know by (3.32), (3.33) where $p = 2m + 2$ that

$$\lim_{\epsilon \to 0} \beta(J_\epsilon(X_\epsilon)) = \beta(X) \quad \text{weakly in } L^{\frac{2m+2}{2m+1}}([0, T] \times \mathcal{O} \times \Omega). \tag{3.42}$$

Now in view of Proposition 3.5.4 we deduce that $\{\beta(J_\epsilon(X_\epsilon))\}$ is in a bounded subset of $L^2(0, T; L^2(\Omega; H^1_0))$ and the conclusion follows from a standard argument. The last part follows from Theorem 3.4.1. $\qquad\qquad\qquad\qquad\qquad\qquad\qquad\qquad\square$

Remark 3.5.6 In the special case of additive noise, that is for $\sigma(X) = \sigma_0$ Eq. (3.1) reduces via the transformation $y = X - \sigma_0 W$ to the random differential equation

$$\begin{cases} \dfrac{\partial y}{\partial t} - \Delta\beta(y + \sigma_0 W(t)) = 0 & \text{in } (0, T) \times \Omega, \\[2mm] \beta(y + \sigma_0 W(t)) = 0 & \text{on } (0, T) \times \partial\Omega \\[2mm] Y(0) = x. \end{cases} \tag{3.43}$$

Though the operator $y \to A(t)y := -\Delta\beta(y + \sigma_0 W(t))$ is maximal monotone in H^{-1} the existence theory for nonlinear Cauchy problems of monotone type (see e.g. [6, 35]) is not applicable because $t \to (I + A(t))^{-1}$ is not of bounded variation on $[0, T]$.

Remark 3.5.7 The previous existence results remain true for time dependent $\beta = \beta(t, r)$ where $\beta \in C([0, T] \times \mathbb{R})$ and as a function of r satisfies Hypothesis 4 or 5 uniformly with respect to $t \in [0, T]$. The argument is exactly the same.

3.6 The Rescaling Approach to Porous Media Equations

In this section we introduce the *rescaling approach*.
   We come back to Eq. (3.1) with

$$\beta(r) = |r|^{m-1}r, \quad \forall\, r \in \mathbb{R}, \ m > 0,$$

and

$$\sigma(x)h = \sum_{k=1}^{N} \mu_k \langle h, e_k \rangle_{-1} x e_k, \quad \forall\, x, h \in H^{-1},$$

where $N \in \mathbb{N}$ and $\mu_1, \ldots, \mu_N \in \mathbb{R}$. Then we have

$$\begin{cases} dX = \Delta(|X|^{m-1}X)dt + \displaystyle\sum_{k=1}^{N} \mu_k(Xe_k)dW_k(t), \quad \text{in } [0,T] \times \mathcal{O}, \\[2mm] X(0) = x \quad \text{in } \mathcal{O}, \\[2mm] X = 0 \quad \text{on } [0,T] \times \partial\mathcal{O} \end{cases} \tag{3.44}$$

Let us consider the Doss–Sussman transformation

$$Y(t) = e^{-\sum_{k=1}^{N} \mu_k e_k W_k(t)} X(t), \tag{3.45}$$

already used for finite dimensional stochastic equations with linear multiplicative noise. Then we see via Itô's formula that if X is a strong solution to (3.44) then Y satisfies the random differential equation

$$\begin{cases} \dfrac{dY}{dt} = e^W \Delta(e^{-W}|Y|^{m-1}Y)dt - \dfrac{1}{2}\,\mu Y, \quad \text{in } [0,T] \times \mathcal{O}, \\[3mm] Y(0) = x =: Y_0, \text{ in } \mathcal{O}, \\[2mm] Y = 0, \quad \text{on } [0,T] \times \partial\mathcal{O}, \end{cases} \tag{3.46}$$

where

$$W(t, \xi) = \sum_{k=1}^{N} \mu_k e_k(\xi) W_k(t)$$

and

$$\mu := \sum_{k=1}^{N} \mu_k^2 e_k^2.$$

Conversely, if Y is a strong solution to (3.46) and $t \to Y(t)$ is an adapted process to the filtration $(\mathscr{F}_t)_{t \geq 0}$, then $X = e^{\sum_{k=1}^{N} \mu_k e_k W_k(t)} Y(t)$ is a strong solution for (3.44). (We refer to [15, 19], for a rigorous proof of the equivalence of (3.44) and (3.46).)

It should be mentioned that (3.46) is not for each $\omega \in \Omega$, the classical standard porous media equation and not a nonlinear parabolic equation either and so its existence can not be reduced to standard existence theory. We have, however, the following

Theorem 3.6.1 *Assume* $1 \le d \le 3$, $m \in (1, 5]$, *and* $Y_0 \in L^{\infty}(\mathcal{O})$. *Then for almost all* $\omega \in \Omega$ *Eq.* (3.46) *has a unique solution* $Y = Y(t, Y_0)$ *satisfying*

(i) $Y \in L^{\infty}((0, T) \times \mathcal{O}) \cap C([0, T]; H^{-1}(\mathcal{O}))$,
(ii) $Y|Y|^{m-1} \in L^2(0, T; H_0^1(\mathcal{O}))$,
(iii) $\frac{dY}{dt} \in L^2(0, T; H^{-1}(\mathcal{O}))$.

Moreover, if $Y_0 \ge 0$ *on* $\mathcal{O}$, *then* $Y \ge 0$ *on* $(0, T) \times \mathcal{O}$.

Here $\frac{d}{dt}$ is the strong derivative of $Y : [0, T] \to H^1(\mathcal{O})$.

In the case $m \in (0, 1)$ which corresponds to the fast diffusion porous media equation, we have

Theorem 3.6.2 *Assume* $1 \le d \le 3$, $0 < m \le 1$, *and* $m \ge \frac{1}{5}$, *if* $d = 3$. *Then for each* $Y_0 \in L^{m+1}(\mathcal{O})$ *Eq.* (3.46) *has a solution* $Y = Y(t, Y_0)$ *satisfying*

(i) $Y \in C([0, T]; H^{-1}(\mathcal{O}))$,
(ii) $Y|Y|^{m-1} \in L^2(0, T; H_0^1(\mathcal{O}))$,
(iii) $\frac{dY}{dt} \in L^2(0, T; H^{-1}(\mathcal{O}))$.

Moreover, if $Y_0 \ge 0$ *on* $\mathcal{O}$, *then* $Y \ge 0$ *on* $(0, T) \times \mathcal{O}$.

For the proofs of Theorems 3.6.1 and 3.6.2 which involve a sharp analysis of Eq. (3.44) based on approximation of β_j by a smooth family of processes $\{\beta_j^{\epsilon}\}_{\epsilon > 0}$ we refer to [15].

Theorems 3.6.1 and 3.6.2 can be used to give a direct proof of existence for equation (3.1). On the other hand, in many situations (see for instance Sects. 3.7 and 3.9 below) the random equation (3.46) can be used to obtain sharp pointwise estimates for the solution X of (3.1).

3.7 Extinction in Finite Time for Fast Diffusions and Self Organized Criticality

We are here concerned with Eq. (3.1) with $\beta(x) = \rho|x|^{\alpha} \operatorname{sign} x$, for some $\alpha \in [0, 1)$ and $\rho > 0$. We have proved that, under the Hypothesis 4 (3.1) has a unique distributional solution $X(t)$, $t \ge 0$ (see Definition 3.1.2 and Theorem 3.4.1.)

In this section we are going to show that, under the more stringent Hypothesis 6 below, there is a finite stopping time τ such that $X(t) = 0$ for $t \ge \tau$ with a finite probability called *extinction probability*.

Hypothesis 6

(i) $\beta(x) = \rho|x|^\alpha \, sign \, x, \quad \alpha \in [0,1), \, \rho > 0.$
(ii) *Hypothesis 4(ii),(iii),(v) hold.*
(iii) *There exists $M_3' > 0$ such that*

$$\|\sigma(x)\|^2_{\mathcal{L}_2(H^{-1})} \leq M_3'\|x\|^2_{-1}, \quad \forall \, x \in H^{-1}. \tag{3.47}$$

(iv) *We have*

$$1 \leq d < \frac{2(1+\alpha)}{1-\alpha}. \tag{3.48}$$

Note that (iii) is more restrictive than Hypothesis 4(iv). By Theorem 3.4.1 Eq. (3.1) has a unique distributional solution X.

Note that when $\alpha = 0$ that is in the case of *stochastic self-organized criticality*, Eq. (3.48) implies $d = 1$.

Example 3.7.1 Assume that $\sigma(x) = kxA^{-\gamma/2}$ with $\gamma > \frac{3}{2}d$, then (iii) holds, see Example 2.1.2).

In order to prove extinction, we need an estimate of $\|X_\epsilon(t)\|^{1-\alpha}_{-1}$, where X_ϵ is the solution to the approximating problem (3.10).

Lemma 3.7.2 *Assume that Hypothesis 6 is fulfilled. Then for any $\epsilon > 0$ and $0 \leq r \leq t$ we have*

$$\|X_\epsilon(t)\|^{1-\alpha}_{-1} + \frac{1-\alpha}{1+\alpha}\int_r^t e^{M_{3,\alpha}(t-s)}\|X_\epsilon(s)\|^{-1-\alpha}_{-1}|X_\epsilon(s)|^{1+\alpha}_{1+\alpha}\mathbb{1}_{\|X_\epsilon(s)\|_{-1}>0}\,ds$$

$$\leq e^{M_{3,\alpha}(t-r)}\|X_\epsilon(r)\|^{1-\alpha}_{-1}$$

$$+(1+\alpha)\int_r^t e^{M_{3,\alpha}(t-s)}\|X_\epsilon(s)\|^{-1-\alpha}_{-1}\langle X_\epsilon(s), \sigma(X_\epsilon(s))\mathbb{1}_{\|X_\epsilon(s)\|_{-1}>0}\,dW(s)\rangle_{-1} \tag{3.49}$$

Proof For the sake of simplicity we take $\rho = 1$. We start by estimating $\|X_\epsilon(t)\|^{1-\alpha}_{-1}$ and to this aim we shall first estimate $\phi_\lambda(X_\epsilon(t))$ where $\phi_\lambda(x) = (\|x\|^2_{-1} + \lambda^2)^{\frac{1-\alpha}{2}}$, $\lambda > 0$.

Notice that

$$D\phi_\lambda(x) = (1-\alpha)(\|x\|^2_{-1} + \lambda^2)^{-\frac{1+\alpha}{2}}x, \tag{3.50}$$

and

$$D^2\phi_\lambda(x) = (1-\alpha)(\|x\|^2_{-1} + \lambda^2)^{-\frac{1+\alpha}{2}} - (1-\alpha^2)(\|x\|^2_{-1} + \lambda^2)^{-\frac{3+\alpha}{2}}x \otimes x. \tag{3.51}$$

By Itô's formula we have

$$d\phi_\lambda(X_\epsilon(t)) + (1-\alpha)(\|X_\epsilon(t)\|_{-1}^2 + \lambda^2)^{-\frac{1+\alpha}{2}} \langle X_\epsilon(t), \beta_\epsilon(X_\epsilon(t))\rangle_2$$

$$= (1+\alpha)(\|X_\epsilon(t)\|_{-1}^2 + \lambda^2)^{-\frac{1+\alpha}{2}} \langle X_\epsilon(t), \sigma(X_\epsilon(t))dW(t)\rangle_{-1}$$

$$+ \tfrac{1-\alpha}{2} (\|X_\epsilon(t)\|_{-1}^2 + \lambda^2)^{-\frac{1+\alpha}{2}} \operatorname{Tr}[\sigma(X_\epsilon(t))\sigma^*(X_\epsilon(t))]dt$$

$$- \tfrac{1-\alpha^2}{2} (\|X_\epsilon(t)\|_{-1}^2 + \lambda^2)^{-\frac{3+\alpha}{2}} dt.$$

Integrating with respect to t from r to t, neglecting the last term and taking into account (3.51), yields

$$\phi_\lambda(X_\epsilon(t)) + \frac{1-\alpha}{1+\alpha} \int_r^t (\|X_\epsilon(s)\|_{-1}^2 + \lambda^2)^{-\frac{1+\alpha}{2}} |X_\epsilon(s)|_{1+\alpha}^{1+\alpha} ds$$

$$\le \phi_\lambda(X_\epsilon(r)) + (1+\alpha) \int_r^t (\|X_\epsilon(s)\|_{-1}^2 + \lambda^2)^{-\frac{1+\alpha}{2}} \langle X_\epsilon(s), \sigma(X_\epsilon(s))dW(s)\rangle_{-1}$$

$$+ \frac{1-\alpha}{2} \int_r^t (\|X_\epsilon(s)\|_{-1}^2 + \lambda^2)^{-\frac{1+\alpha}{2}} \operatorname{Tr}[\sigma(X_\epsilon(s))\sigma^*(X_\epsilon(s))]ds.$$

Letting $\lambda \to 0$, yields

$$\|X_c(t)\|_{-1}^{1-\alpha} + \frac{1-\alpha}{1+\alpha} \int_r^t \|X_c(s)\|_{-1}^{-1-\alpha} |X_c(s)|_{L^{1+\alpha}}^{1+\alpha} \mathbb{1}_{\|X_\epsilon(s)\|_{-1}>0} ds$$

$$\le \|X_\epsilon(r)\|_{-1}^{1-\alpha} + (1+\alpha) \int_r^t \|X_\epsilon(s)\|_{-1}^{-1-\alpha} \langle X_\epsilon(s), \sigma(X_\epsilon(s)) \mathbb{1}_{\|X_\epsilon(s)\|_{-1}>0} dW(s)\rangle_{-1}$$

$$+ \frac{1-\alpha}{2} \int_r^t \|X_\epsilon(s)\|_{-1}^{-1-\alpha} \operatorname{Tr}[\sigma(X_\epsilon(s))\sigma^*(X_\epsilon(s))] \mathbb{1}_{\|X_\epsilon(s)\|_{-1}>0} ds.$$

Taking into account (3.47) we find

$$\|X_\epsilon(t)\|_{-1}^{1-\alpha} + \frac{1-\alpha}{1+\alpha} \int_r^t \|X_\epsilon(s)\|_{-1}^{-1-\alpha} |X_\epsilon(s)|_{1+\alpha}^{1+\alpha} \mathbb{1}_{\|X_\epsilon(s)\|_{-1}>0} ds$$

$$\le \|X_\epsilon(r)\|_{-1}^{1-\alpha} + (1+\alpha) \int_r^t \|X_\epsilon(s)\|_{-1}^{-1-\alpha} \langle X_\epsilon(s), \sigma(X_\epsilon(s)) \mathbb{1}_{\|X_\epsilon(s)\|_{-1}>0} dW(s)\rangle_{-1}$$

$$+ \frac{M_3(1-\alpha)}{2} \int_r^t \|X_\epsilon(s)\|_{-1}^{1-\alpha} ds.$$

$$(3.52)$$

Finally, from the stochastic Gronwall lemma we get, setting $M_{3,\alpha} := \frac{1-\alpha}{2} M_3$

$$\|X_\epsilon(t)\|_{-1}^{1-\alpha} + \frac{1-\alpha}{1+\alpha} \int_r^t e^{M_{3,\alpha}(t-s)} \|X_\epsilon(s)\|_{-1}^{-1-\alpha} |X_\epsilon(s)|_{1+\alpha}^{1+\alpha} \, \mathbb{1}_{\|X_\epsilon(s)\|_{-1}>0} \, ds$$

$$\leq e^{M_{3,\alpha}(t-r)} \|X_\epsilon(r)\|_{-1}^{1-\alpha}$$

$$+(1+\alpha) \int_r^t e^{M_{3,\alpha}(t-s)} \|X_\epsilon(s)\|_{-1}^{-1-\alpha} \langle X_\epsilon(s), \sigma(X_\epsilon(s)) \, \mathbb{1}_{\|X_\epsilon(s)\|_{-1}>0} \, dW(s)\rangle_{-1}.$$
$$(3.53)$$

Therefore (3.49) is fulfilled.   □

Now we are ready to study the extinction in finite time of $X(t)$. For this we need an additional assumption

$$L^{1+\alpha} \subset H^{-1} \text{ with continuous embedding.} \tag{3.54}$$

This is equivalent to the existence of $c_\alpha > 0$ such that

$$\|x\|_{-1}^{-1-\alpha} |x|_{1+\alpha}^{1+\alpha} \geq c_\alpha, \quad \forall \, x \in H^{-1}. \tag{3.55}$$

Condition (3.54) implies a restriction on the dimension d of the space. In fact by duality it is equivalent to

$$H_0^1 \subset L^{\frac{1+\alpha}{\alpha}} \text{ with continuous embedding.} \tag{3.56}$$

Now by the Sobolev embedding theorem (3.54) holds provided

$$\frac{\alpha}{1+\alpha} > \frac{1}{2} - \frac{1}{d},$$

which is just condition (3.48).

Theorem 3.7.3 *Assume, besides Hypothesis 6 that (3.48) is fulfilled. Let $X(t)$ be a generalized solution to problem (3.1) and set*

$$\tau := \inf\{t > 0 : X(t) = 0\}.$$

Then

$$X(t,\omega) = 0, \quad \forall t > \tau(\omega).$$

(we say that process $X(t)$ extinguishes after time τ). Moreover the extinction probability is finite and

$$\mathbb{P}(\tau_x > t) \leq \frac{\|x\|_{-1}^{1-\alpha}}{\frac{c_\alpha(1-\alpha)}{M_{3,\alpha}(1+\alpha)}(1 - e^{-M_{3,\alpha}t})} .$$

(3.57)

Proof By (3.49), taking into account (3.55), we have for all $0 \leq r \leq t$

$$\|X_\epsilon(t)\|_{-1}^{1-\alpha} + c_\alpha \frac{1-\alpha}{1+\alpha} \int_r^t e^{M_{3,\alpha}(t-s)} \mathbb{1}_{\|X_\epsilon(s)\|_{-1}>0} ds$$

$$\leq e^{M_{3,\alpha}(t-r)}\|X_\epsilon(r)\|_{-1}^{1-\alpha} + (1+\alpha)$$

(3.58)

$$\times \int_r^t e^{M_{3,\alpha}(t-s)}\|X_\epsilon(s)\|_{-1}^{-1-\alpha} \langle X_\epsilon(s), \sigma(X_\epsilon(s))\mathbb{1}_{\|X_\epsilon(s)\|_{-1}>0} dW(s)\rangle_{-1}$$

Finally, letting $\epsilon \to 0$ we find

$$\|X(t)\|_{-1}^{1-\alpha} + c_\alpha \frac{1-\alpha}{1+\alpha} \int_r^t e^{M_{3,\alpha}(t-s)} \mathbb{1}_{\|X(s)\|_{-1}>0} ds$$

$$\leq e^{M_{3,\alpha}(t-r)}\|X_\epsilon(r)\|_{-1}^{1-\alpha} + (1+\alpha)$$

(3.59)

$$\times \int_r^t e^{M_{3,\alpha}(t-s)}\|X(s)\|_{-1}^{-1-\alpha} \langle X(s), \sigma(X_\epsilon(s))\mathbb{1}_{\|X(s)\|_{-1}>0} dW(s)\rangle_{-1}.$$

Setting now $Y(t) = e^{-M_{3,\alpha}(t-r)}X(t)$ we deduce by (3.59) that

$$\|Y(t)\|_{-1}^{1-\alpha} + c_\alpha \frac{1-\alpha}{1+\alpha} \int_r^t e^{-M_{3,\alpha}s}\mathbb{1}_{\|X(s)\|_{-1}>0} ds \leq e^{-M_{3,\alpha}(t-r)}\|Y(r)\|_{-1}^{1-\alpha}$$

$$+(1+\alpha) \int_r^t e^{-M_{3,\alpha}s}\|X(s)\|_{-1}^{-1-\alpha} \langle X(s), \sigma(X_\epsilon(s))\mathbb{1}_{\|X(s)\|_{-1}>0} dW(s)\rangle_{-1},$$

(3.60)

for all $0 \leq r \leq t$. It follows therefore that $(\|Y(t)\|_{-1}^{1-\alpha})_{t\geq 0}$ is a nonnegative super-martingale, that is,

$$\mathbb{E}\left[\|Y(t)\|_{-1}^{1-\alpha} | \mathscr{F}_r\right] \leq \|Y(r)\|_{-1}^{1-\alpha}, \quad \forall \, t \geq r.$$

This implies for any couple of stopping times τ, σ that

$$\tau > \sigma \Rightarrow Y(\tau) \leq Y(\sigma).$$

In particular, for any $t > \tau$ where

$$\tau = \inf\{t > 0 : X(t) = 0\},$$

we have

$$Y(t) \le Y(\tau) = 0, \text{ equivalently } X(t) = X(\tau) = 0, \quad \mathbb{P}\text{-a.s.}.$$

So, the extinction occurs at the moment τ.

Let us finally estimate of $\mathbb{E}(\tau < \infty)$. Taking expectation in (3.60) with $r = 0$, yields

$$\mathbb{E}\|Y(t)\|_{-1}^{1-\alpha} + c_\alpha \frac{1-\alpha}{1+\alpha} \int_0^t e^{-M_{3,\alpha}s} \mathbb{P}(\tau > s)\, ds \le \|x\|_{-1}, \qquad (3.61)$$

from which

$$\mathbb{E}\|Y(t)\|_{-1}^{1-\alpha} + c_\alpha \frac{1-\alpha}{1+\alpha} \int_0^t e^{-M_{3,\alpha}s} \mathbb{P}(\tau > t)\, ds \le \|x\|_{-1}^{1-\alpha}, \qquad (3.62)$$

Therefore

$$\frac{c_\alpha(1-\alpha)}{M_{3,\alpha}(1+\alpha)}(1 - e^{-M_{3,\alpha}t})\mathbb{P}(\tau_x > t) \le \|x\|_{-1}^{1-\alpha}$$

and (3.58) follows. □

3.8　The Asymptotic Extinction of Solutions to Self Organized Criticality

We consider here the self organized criticality stochastic equation (1.8), that is

$$\begin{cases} dX = \rho\Delta(\text{sign } X)dt + \sum_{k=1}^{\infty} \mu_k X e_k\, dW_k(t), & \text{in } [0, T] \times \mathscr{O}, \\[2ex] X(0) = x, & \text{on } \mathscr{O}, \\[2ex] \text{sign } X \ni 0, & \text{on } [0, T] \times \partial\mathscr{O}, \end{cases} \qquad (3.63)$$

where $\mu_i \in \mathbb{R}$ are chosen in such a way that

$$\sum_{k=1}^{\infty} \mu_k \lambda_k^2 < \infty$$

and λ_k are the eigenvalues corresponding to eigenvectors e_k of $-\Delta$ (with the Dirichlet boundary conditions).

By Theorem 3.4.1 we know that for each $T > 0$ and all $x \in L^p$, $p \geq 2$, problem (3.63) has a unique distributional solution X which is nonnegative whenever x is nonnegative.

Moreover, we know by Theorem 3.6.2 that for $d = 1$ we have finite extinction with probability given by (3.57). Here we shall study the asymptotic behaviour in the case $d > 1$.

Theorem 3.8.1 *Assume* $x \in L_+^4$. *The solution X to Eq. (3.1) satisfies*

$$(i) \quad \lim_{t \to \infty} \int_{\mathcal{O}} X(t, \xi) d\xi = l < \infty, \quad \mathbb{P}\text{-}a.s., \qquad (3.64)$$

$$(ii) \quad \int_0^\infty m(\mathcal{O} \setminus \mathcal{O}_0^t) dt < \infty, \quad \mathbb{P}\text{-}a.s., \qquad (3.65)$$

where m is the Lebesgue measure and

$$\mathcal{O}_0^t = \{\xi \in \mathcal{O} : X(t, \xi) = 0\}, \quad t \geq 0. \qquad (3.66)$$

By (3.66) it follows that for *almost all* sequences $t_n \to \infty$ we have $m(\mathcal{O} \setminus \mathcal{O}_0^{t_n}) \to 0$. Roughly speaking this means that for t large enough $X(t, \xi) = 0$ on a set $\mathcal{O}_0^t$ which differs from $\mathcal{O}$ by a set of small Lebesgue measure. In other words, for t large enough the *non critical zone* $\mathcal{O} \setminus \mathcal{O}_0^t$ of $X(t)$ is *arbitrarily small*. Equation (3.64) means that the total mass associated with the process $X(t)$ is $\mathbb{P}$-a.s. convergent as $t \to \infty$. One might suspect that $l = 0$ (as it happens in deterministic case [7]) and we shall see that this is indeed the case for a special form of the Wiener process $W(t)$. (In this direction a sharper result was obtained recently in [61].)

Proof Since the complete proof of the theorem is given in the work [17] here we confine ourselves to sketch it and we refer to the above mentioned paper for details.

We come back to the approximating Eq. (3.10) where $\beta = \rho \text{sgn}$ and note that it follows via a standard martingale integral inequality that for each $T > 0$

$$\mathbb{E} \sup_{t \in [0,T]} |X(t)|_2^2 \leq C_T |x|_2^2. \qquad (3.67)$$

(The details are omitted.)

Next we consider a function $\varphi_\lambda \in C_b^3(\mathbb{R})$ such that $\varphi_\lambda(0) = 0$ and

$$\begin{cases} \varphi_\lambda'(r) = \frac{r}{\lambda} \text{ for } |r| \leq \lambda, \quad \varphi_\lambda'(r) = 1 + \lambda \text{ for } r \geq 2\lambda \\ \\ \varphi_\lambda'(r) = -(1 + \lambda) \text{ for } r \leq -2\lambda, \quad |\varphi_\lambda''(r)| \leq \frac{C}{\lambda} \text{ for } |r| \leq 2\lambda, \end{cases} \qquad (3.68)$$

for some $C > 0$. (We may choose $\psi_\lambda = \varphi_\lambda'$ of the form $ar^4 + br^3 + cr^2 + dr$ where a, b, c, d are determined by the conditions $\psi_\lambda(\lambda) = 1$, $\psi_\lambda'(\lambda) = \frac{1}{\lambda}$, $\psi_\lambda(2\lambda) = 1 + 2\lambda$, $\psi_\lambda'(2\lambda) = 0$).)

It is easily seen that φ_λ is a smooth approximation of the function $r \mapsto |r|$ and

$$|\varphi_\lambda'(r) - (\text{sign})_\lambda(r)| \leq C\lambda, \quad \forall\, r \in \mathbb{R}, \ \lambda > 0, \tag{3.69}$$

where $(\text{sign})_\lambda$ is the Yosida approximation of the sign graph, i.e. $\beta_\lambda = \rho(\text{sign})_\lambda$.

Next we set $Y_\lambda^\epsilon := (1 + \epsilon A)^{-1} X_\lambda$, where $A = -\Delta$, $D(A) := H^2 \cap H_0^1$, $\epsilon > 0$ and rewrite (3.10) in terms of Y_λ^ϵ. We obtain that

$$\begin{cases} dY_\lambda^\epsilon + A(1 + \epsilon A)^{-1}(\beta_\lambda(X_\lambda) + \lambda X_\lambda)dt = (1 + \epsilon A)^{-1} X_\lambda dW(t), & \text{in } (0, \infty) \times \mathcal{O}, \\[2mm] \beta_\lambda(X_\lambda)) + \lambda X_\lambda = 0, & \text{on } (0, \infty) \times \partial\mathcal{O}, \\[2mm] Y_\lambda^\epsilon(0) = (1 + \epsilon A)^{-1} x, & \text{in } \mathcal{O}. \end{cases} \tag{3.70}$$

The process $t \mapsto Y_\lambda^\epsilon(t)$ is H_0^1-valued and continuous on $[0, T]$ and so, applying Itô's formula in (3.70) and letting $\epsilon \to 0$, yields

$$\int_\mathcal{O} \varphi_\lambda(X_\lambda)d\xi + \int_0^t \int_\mathcal{O} \nabla(\beta_\lambda(X_\lambda) + \lambda X_\lambda) \cdot \nabla\varphi_\lambda'(X_\lambda)ds\, d\xi$$

$$= \int_\mathcal{O} \varphi_\lambda(x)d\xi + \sum_{k=1}^\infty \mu_k^2 \int_0^t \int_\mathcal{O} \varphi_\lambda''(X_\lambda)|(X_\lambda e_k)|^2 d\xi\, ds \tag{3.71}$$

$$+ \int_0^t \langle \varphi_\lambda'(X_\lambda), X_\lambda dW(s) \rangle_2.$$

We also note that by (3.68) we have

$$\sum_{k=1}^\infty \mu_k^2 \int_0^t \int_\mathcal{O} \varphi_\lambda''(X_\lambda)|(X_\lambda e_k)|^2 d\xi\, ds$$

$$\tag{3.72}$$

$$\leq 4C\lambda \sum_{k=1}^\infty \mu_k^2 \int_0^t \int_\mathcal{O} \mathbb{1}_\lambda(s, \xi)d\xi\, ds,$$

where $\mathbb{1}_\lambda$ is the characteristic function of the set

$$\{(s, \xi, \omega) \in (0, \infty) \times \mathcal{O} \times \Omega : 0 \leq X_\lambda(s, \xi, \omega) \leq 2\lambda\}.$$

It follows also that

$$\lim_{\lambda \to 0} \int_{\mathcal{O}} \varphi_\lambda(X_\lambda(t, \xi)) d\xi = \int_{\mathcal{O}} X(t, \xi) d\xi, \text{ weakly in } L^2(\Omega), \ \forall \, t \geq 0. \qquad (3.73)$$

We set

$$I_\lambda(t) = \int_0^t \int_{\mathcal{O}} \nabla(\beta_\lambda(X_\lambda) + \lambda X_\lambda) \cdot \nabla \varphi_\lambda'(X_\lambda) d\xi \, ds,$$

$$M_\lambda(t) = \int_0^t \langle \varphi_\lambda'(X_\lambda), X_\lambda) dW \rangle_2 = \sum_{k=1}^{\infty} \int_0^t \langle \varphi_\lambda'(X_\lambda), X_\lambda e_k) d\beta_k(s) \rangle_2$$

and so we rewrite (3.71) as

$$\int_{\mathcal{O}} \varphi_\lambda(X_\lambda(t)) d\xi + I_\lambda(t)$$

$$\qquad\qquad\qquad\qquad\qquad\qquad\qquad\qquad\qquad\qquad (3.74)$$

$$= \int_{\mathcal{O}} \varphi_\lambda(x) d\xi + \sum_{k=1}^{\infty} \mu_k^2 \int_0^t \int_{\mathcal{O}} \varphi_\lambda''(X_\lambda(t)) |(X_\lambda(t) e_k)|^2 d\xi \, ds + M_\lambda(t).$$

Taking into account that

$$X_\lambda \to X, \quad \varphi_\lambda'(X_\lambda) \to \eta \in \rho \, \text{sign} \, X, \text{ weakly in } L^2((0, \infty) \times \mathcal{O} \times \Omega),$$

it follows after some calculations that $\mathbb{P}$-a.s.

$$\lim_{\lambda \to 0} M_\lambda(t) = M(t) = \int_0^t \langle \eta, X(s) dW(s) \rangle_2, \quad \forall \, t \geq 0. \qquad (3.75)$$

Then by (3.16)–(3.21) we see that

$$\int_{\mathcal{O}} \varphi(X(t, \xi)) d\xi + \widetilde{I}(t) = \int_{\mathcal{O}} \varphi(x) d\xi + M(t), \quad \forall \, t \geq 0, \qquad (3.76)$$

where

$$\widetilde{I}(t) = w - \lim_{\lambda \to 0} I_\lambda(t) \quad \text{in } L^2(\Omega). \qquad (3.77)$$

We set

$$Z(t) = \int_{\mathcal{O}} \varphi(X(t, \xi)) d\xi$$

and note that it is a nonnegative semimartingale with $\mathbb{E}Z(t) < \infty, \forall\, t \geq 0$. Since the function $t \mapsto X(t)$ is a weakly continuous L^2-valued function it follows also that $t \mapsto Z(t)$ is continuous. Then we may define a continuous version $I(t)$ of $\widetilde{I}(t)$

$$I(t) = Z(0) - Z(t) + M(t), \quad \forall\, t \geq 0 \tag{3.78}$$

and it follows also that I is a nondecreasing process on $(0, \infty)$. Moreover $M(t)$ is a continuous semimartingale.

Applying Lemma 1.2.4 to (3.78) we infer that

$$\lim_{t \to \infty} \int_{\mathscr{O}} \varphi(X(t,\xi))d\xi = l < \infty,$$

exists $\mathbb{P}$-a.s.

Now coming back to I_λ we see that $\mathbb{P}$-a.s.

$$I_\lambda(t) \geq \int_0^t \int_{\mathscr{O}} \nabla(\beta_\lambda(X_\lambda) + \lambda X_\lambda) \cdot \nabla\varphi_\lambda'(X_\lambda)d\xi\, ds$$
$$\geq \int_0^t \int_{\mathscr{O}} |\nabla\beta_\lambda(X_\lambda)|^2 d\xi\, ds.$$

Taking into account that $\Psi_\lambda(X_\lambda) \to \eta \in \rho\,\mathrm{sign}\, X$ weakly in $L^2((0,\infty)\times\mathscr{O}\times\Omega)$ as $\lambda \to 0$ we infer that

$$\int_0^t |\nabla\eta|_2^2\, dt \leq I(t), \quad t \geq 0,\ \mathbb{P}\text{-a.s.}$$

and therefore

$$\int_0^\infty |\nabla\eta|_2^2\, dt \leq I(\infty), \quad \mathbb{P}\text{-a.s.}$$

Next by the Sobolev embedding theorem we have

$$|\eta(t)|_{p^*} \leq C|\nabla\eta|_2, \quad \forall\, t \geq 0,$$

where $p^* = \frac{2d}{d-2}$ for $d > 2$, p^* arbitrary in $[2,\infty)$ for $d=2$ and $p^* = \infty$ for $d=1$. Hence

$$\int_0^\infty |\eta|_{p^*}^2\, dt \leq \infty, \quad t \geq 0,\ \mathbb{P}\text{-a.s.} \tag{3.79}$$

Taking into account that $\eta \in \rho\,\mathrm{sign}\, X$ a.e. in $(0,\infty)\times\mathscr{O}\times\Omega$, we have $\eta = \rho$ a.e. in $\{(t,\xi,\omega) :\ X((t,\xi,\omega)) > 0\}$ and so (3.79) yields

$$\int_0^\infty (m(\mathscr{O}\setminus\mathscr{O}_0^t))^{\frac{2}{p^*}}\, dt < \infty,$$

and we get (3.65) as claimed.   □

We shall now assume that the noise is finite dimensional, that is

$$W(t, \xi) = \sum_{k=1}^{N} \mu_k e_k(\xi) W_k(t), \quad t \geq 0, \; \xi \in \mathcal{O}, \tag{3.80}$$

and set

$$\mu(\xi) = \sum_{k=1}^{N} \mu_k^2 e_k^2(\xi), \quad \xi \in \mathcal{O}. \tag{3.81}$$

In this case Theorem 3.8.1 is completed by the following asymptotic result.

Theorem 3.8.2 *Under the assumptions of Theorem 3.8.1 assume further that W is of the form (3.80). Then we have*

$$\lim_{t \to \infty} e^{-W(t)} X(t) = 0 \quad in \; L^1, \; \mathbb{P}\text{-}a.s. \tag{3.82}$$

and if $\mu(\xi) > 0$ for all $\xi \in \mathcal{O}$

$$\lim_{t \to \infty} X(t) = 0 \quad in \; L^1_{loc}, \quad \mathbb{P}\text{-}a.s. \tag{3.83}$$

Moreover, for each compact subset $K \subset \mathcal{O}$ we have

$$\int_K X(t, \xi) d\xi$$

$$\leq (m(K))^{\frac{1}{2}} |x|_2 \exp \left\{ \sup_K (\mu)^{\frac{1}{2}} \left(\sum_{k=1}^{N} |\beta_k(t)| \right)^{\frac{1}{2}} \right\} e^{-\frac{t}{2} \inf_{K'} \mu}, \tag{3.84}$$

where K' is any compact neighborhood of K. In particular, one has

$$\int_K X(t, \xi, \omega) d\xi \leq (m(K))^{\frac{1}{2}} |x|_2 e^{-\rho_K t}, \quad \forall \, t \geq t_0(\omega), \; \omega \in \Omega, \tag{3.85}$$

for some $\rho_K > 0$.

It should be noted that the condition $\mu > 0$ on $\mathcal{O}$ automatically holds if $\mu_1 > 0$ because the first eigenfunction e_1 of the Laplace operator with homogeneous boundary conditions is positive on $\mathcal{O}$.

Proof We proceed as in Sect. 3.6 by rescaling Eq. (1.1) via the Doss–Sussman transformation

$$X(t) = e^{W(t)} Y(t)$$

and so to reduce it to the random differential equation

$$\begin{cases} \dfrac{\partial Y(t)}{\partial t} - e^{-W(t)} \Delta \beta (e^{W(t)} Y(t)) + \dfrac{1}{2} \mu \, Y(t) = 0 \quad \text{in } (0,\infty) \times \mathcal{O} \\[2mm] \beta(e^{W(t)} Y(t))) \in H_0^1, \quad \forall \, t \geq 0, \ \mathbb{P}\text{-a.s.,} \\[2mm] Y(0) = x. \end{cases} \tag{3.86}$$

(Here $\frac{\partial Y}{\partial t}$ is taken in H^{-1}.)

We first note that via the regularized equation we have that

$$|Y(t)|_2 \leq |x|_2, \quad \forall \ \mathbb{P}\text{-a.s..} \tag{3.87}$$

To prove this we consider the solution Y_λ to approximating equation

$$\begin{cases} \dfrac{\partial Y_\lambda}{\partial t} - e^{-W} \Delta (\Psi_\lambda (e^W Y_\lambda) + \lambda e^W Y_\lambda)) + \dfrac{1}{2} \widetilde{\mu} \, Y_\lambda = 0 \quad \text{in } (0,\infty) \times \mathcal{O} \\[2mm] Y_\lambda(0) = x \quad \text{in } \mathcal{O}. \end{cases} \tag{3.88}$$

and get appropriate estimates.

Now let us prove (3.82). Assume that this is not true, that is, there exists $\delta > 0$ such that for some $\{t_n\} \to \infty$

$$\|Y(t_n)\|_1 \geq \delta > 0, \quad \forall \, n \in \mathbb{N}, \tag{3.89}$$

where $Y = Y(t, \omega)$ and $\omega \in \Omega$ is arbitrary but fixed. By estimate (3.87) it follows that there is $f \in L^2$ such that $Y(t_n) \to f$ weakly in L^2 (possibly on a subsequence of $\{t_n\}$.) Clearly by (3.89) we have

$$0 < \delta \leq \int_{\mathcal{O}} f(\xi) d\xi, \quad f \geq 0 \text{ a.e. in } \mathcal{O}$$

and so $f \neq 0$. On the other hand, for each $n \in \mathbb{N}$ there is $t_n > 0$ such that

$$\left| \int_{\mathcal{O}} Y(t, \xi) d\xi - \int_{\mathcal{O}} Y(t_n, \xi) d\xi \right| \leq \frac{1}{n}, \quad \forall \, t \in (t_n - \epsilon_n, t_n + \epsilon_n). \tag{3.90}$$

By (3.5) it follows that there is a subsequence $\{t_{n_k}\} \to \infty$ and $s_k \in (t_{n_k} - \epsilon_{n_k}, t_{n_k} + \epsilon_{n_k})$ such that

$$\int_{\mathcal{O}} \mathbb{1}_{\{X(s_k) \neq 0\}} \, d\xi = m(\mathcal{O} \setminus \mathcal{O}_0^{s_k}) \to 0$$

as $k \to \infty$. Hence (selecting a further subsequence if necessary) we have

$$\mathbb{1}_{\{X(s_k) \neq 0\}} \to 0, \quad \text{a.s. as } k \to \infty.$$

Once again by (3.87) we have that $X(s_k) \to \tilde{f}$ weakly in L^2 and this clearly implies that $Y(s_k) = Y(s_k) \mathbb{1}_{\{X(s_k) \neq 0\}} \to 0$ a.e. as $k \to \infty$ and so $\tilde{f} = 0$ a.e. This yields (see (3.90))

$$\int_{\mathcal{O}} f(\xi) \, d\xi = \lim_{k \to \infty} \int_{\mathcal{O}} Y(t_{n_k}, \xi) \, d\xi = \int_{\mathcal{O}} \tilde{f}(\xi) \, d\xi = 0.$$

This contradiction proves (3.82).

To prove (3.84) we consider a compact $K \subset \mathcal{O}$ and $K' \subset \mathcal{O}$ a compact neighborhood of K. Choose a function $\mu^\alpha \in C_0^\infty(\mathcal{O})$ such that $0 \leq \mu^\alpha \leq 1$, $\mu^\alpha \leq 1$ on K and $\mu^\alpha = 0$ on $\mathcal{O} \setminus K$. We set $C_K = \inf_{K'} \tilde{\mu}$.

Multiplying (3.88) by $\mu^\alpha Y_\lambda$ and integrating over $\mathcal{O}$, we obtain after some calculation, that

$$(\mu^\alpha)^{\frac{1}{2}} |Y_\lambda(t)|_2^2 \leq |(\mu^\alpha)^{\frac{1}{2}} x|_2^2 e^{-C_K t} + \lambda \int_0^t e^{-C_K(t-s)} \eta_\lambda(s) ds, \quad \forall \, t \geq 0 \; \mathbb{P}\text{-a.s.}$$

(3.91)

Then letting $\lambda \to 0$ in (3.91) we get

$$|(\mu^\alpha)^{\frac{1}{2}} Y_\lambda(t)|_2^2 \leq e^{-C_K t} \mu^\alpha |x|_2^2 \leq e^{-C_K t} |x|_2^2, \quad \forall \, t \geq 0 \; \mathbb{P}\text{-a.s..}$$

(3.92)

Taking into account that

$$\int_K X(t, \xi) d\xi = \int_K Y(t, \xi) e^{-W(t)} d\xi, \quad \forall \, t \geq 0 \; \mathbb{P}\text{-a.s.,}$$

by (3.90) we obtain the desired estimate (3.84) as claimed. □

Remark 3.8.3 It should be noted that if W is the of form (3.80), but $e_k \in C^2(\overline{\mathcal{O}})$ are such that $|e_k|_\infty > 0$, then $\inf\{\mu(\xi) : \xi \in \mathcal{O}\} > 0$ and so in (3.85) we may replace K by $\overline{\mathcal{O}}$ and so (3.85) implies that

$$\lim_{t \to \infty} \int_{\mathcal{O}} X(t, \xi) d\xi = 0, \quad \mathbb{P}\text{-a.s.}$$

and so, in particular

$$\lim_{t \to \infty} X(t, \xi) d\xi = 0, \quad \text{a.e. in } \mathcal{O} \times \Omega.$$

3.9 Localization of Solutions to Stochastic Slow Diffusion Equations: Finite Speed of Propagation

In this section we state the localization of solutions Let $\mathcal{O}$ be a bounded and open domain of $\mathbb{R}^d$, $d = 1, 2, 3$, with smooth boundary $\partial\mathcal{O}$. We come back to the stochastic porous media equation

$$\begin{cases} dX - \Delta(|X|^{m-1}X)dt = \sigma(X)dW_t, & t \geq 0, \\ X = 0 & \text{on } \partial\mathcal{O}, \qquad (3.93) \\ X(0) = x & \text{in } \mathcal{O}, \end{cases}$$

where $m \geq 1$, W_t is a Wiener process in L^2 of the form

$$W(t) = \sum_{k=1}^{N} W_k(t)e_k. \qquad (3.94)$$

$\{W_k\}_{k=1}^{N}$ is a sequence of independent Brownian motions on a filtered probability space $\{\Omega, \mathscr{F}, \mathscr{F}_t, \mathbb{P}\}$ while $\{e_k\}_{k\in N}$ is an orthonormal basis in L^2 and

$$\sigma(X)W_t = \sum_{k=1}^{N} \mu_k Xe_k W_k(t), \qquad (3.95)$$

where $\{\mu_k\}$ is a sequence of nonnegative numbers. It is well known that the solutions to deterministic slow diffusions porous media equations have the finite speed property. We shall see below that this happens in an appropriate sense for the low diffusion stochastic equation (3.93).

We assume that $e_k \in C^2(\overline{\mathcal{O}})$, $1 \leq k \leq N$, and

$$\sum_{k=1}^{N} \mu_k^2 e_k^2(x) \geq \rho > 0, \quad \forall x \in \overline{\mathcal{O}}. \qquad (3.96)$$

We recall that by Theorem 3.5.5 if $x \in L^{m+1}$, then Eq. (3.93) has a unique strong solution X. If $x \geq 0$ a.e. in $\mathcal{O}$, then $X \geq 0$ a.e. in $\Omega \times (0,T) \times \mathcal{O}$ and

$$\mathbb{E}\int_0^T ds \int_{\mathcal{O}} \left|\nabla(|X|^{m-1}X)\right|^2 d\xi + \sup_{t\in[0,T]} \mathbb{E}\int_{\mathcal{O}} |X(t,\xi)|^{m+1}d\xi \qquad (3.97)$$

$$\leq C\int_{\mathcal{O}} |x|^{m+1}d\xi.$$

Everywhere in the sequel, $B_r(\xi_0) \subset \mathcal{O}$ shall denote the open ball $\{\xi : |\xi - \xi_0| < r\}$, and $\Sigma_r(\xi_0) = \{\xi \in \mathbb{R}^d : |\xi - \xi_0| = r\}$ its boundary, and $B_r^c(\xi_0) = \mathcal{O} \setminus B_r(\xi_0)$,

$\xi_0 \in \mathcal{O}$. As mentioned in the introduction, $\mathcal{O}$ is an open and bounded domain of $\mathbb{R}^d$ with smooth boundary $\partial\mathcal{O}$, $d = 1, 2, 3$. Everywhere below, X is the strong solution to equation with initial data x.

Below, we are only concerned with small $T > 0$, so we may assume that $T \leq 1$. Furthermore, for a function $g : [0, 1] \to \mathbb{R}$, we define its α-Hölder norm, $\alpha \in (0, 1)$, by

$$|g|_\alpha := \sup_{\substack{s,t \in [0,1] \\ s \neq t}} \frac{|g(t) - g(s)|}{|t - s|^\alpha}.$$

Let for $\alpha \in \left(0, \frac{1}{2}\right)$

$$\Omega^\alpha_{H,R} = \{\omega \in \Omega : |\beta_k(\omega)|_\alpha \leq R, \ 1 \leq k \leq N\}.$$

Then, $\Omega^\alpha_{H,R} \nearrow \Omega$ as $R \to \infty$ $\mathbb{P}$-a.s.

Now, we are ready to formulate the main result.

Theorem 3.9.1 *Assume that $d = 1, 2, 3$ and $1 < m \leq 5$, and that $x \in L^\infty(\mathcal{O})$, $x \geq 0$, is such that*

$$\text{support}\{x\} \subset B^c_{r_0}(\xi_0), \tag{3.98}$$

where $r_0 > 0$ and $\xi_0 \in \mathcal{O}$. Fix $\alpha \in \left(0, \frac{1}{2}\right)$ and let for $R > 0$

$$\delta(R) := \left(\frac{1}{m+1}\left(\frac{\rho}{2}\right)^{1/2} c_1^{-1} \left(\sum_{k=1}^N |\nabla e_k|_\infty \mu_k\right)^{-1}\right.$$
$$\left. \times \exp\left[\frac{1}{2}(1-m)\left(\frac{1}{2}c_2 + \sum_{k=1}^N |e_k|_\infty \mu_k\right)\right]\right) \wedge 1,$$

where c_1, c_2 (depending on R) are as in Lemma 3.9.2 below and ρ as in (3.96). Define for $T \in (0, 1]$

$$\Omega_T^{\delta(R)} := \left\{ \sup_{t \in [0,T]} |\beta_k(t)| \leq \delta(R) \text{ for all } 1 \leq k \leq N \right\}.$$

Then, for $\omega \in \Omega_T^{\delta(R)} \cap \Omega^\alpha_{H,R}$, there is a decreasing function $r(\cdot, \omega) : [0, T] \to (0, r_0]$, and $t(\omega) \in (0, T]$ such that for all $0 \leq t \leq t(\omega)$,

$$X(t, \omega) = 0 \text{ on } B_{r(t,\omega)}(\xi_0) \supset B_{r(t(\omega),\omega)}(\xi_0), \text{ and}$$
$$X(t, \omega) \not\equiv 0 \text{ on } B^c_{r(t,\omega)} \subset B^c_{r(t(\omega),\omega)}(\xi_0). \tag{3.99}$$

Since $\Omega_T^{\delta(R)} \nearrow \Omega$ as $T \to 0$ up to a $\mathbb{P}$-zero set, and hence

$$\mathbb{P}\left(\bigcup_{m \in \mathbb{N}} \bigcup_{n \in \mathbb{N}} \Omega_{1/n}^{\delta(m)} \cap \Omega_{H,m}^{\alpha}\right) = 1,$$

it follows that we have finite speed of propagation of disturbances ("localization") for $(X_t)_{t \geq 0}$ $\mathbb{P}$-a.s.

As explicitly follows from the proof, the function $t \to r(t)$ is a process adapted to the filtration $\{\mathscr{F}_t\}$.

Roughly speaking, Theorem 3.9.1 amounts to saying that, for $\omega \in \Omega_T^{\delta(R)} \cap \Omega_{H,R}^{\alpha}$ and for a time interval $[0, t(\omega)]$ sufficiently small, the stochastic flow $X = X(t, \xi, \omega)$ propagates with finite speed.

If we set $r^T(\omega) = \lim_{t \to T} r(t, \omega)$, we see by (3.99) that $X(t, \omega) = 0$ on $B_{r^T(\omega)}$, $\forall t \in (0, t(\omega))$ and $X(t) \not\equiv 0$ on $B_{r^T(\omega)}^c$. It is not clear whether $r^T(\omega) = 0$ for some $T > 0$, that is, whether the "hole filling" property holds in this case (see [89]).

It should be mentioned also that the assumption $x \geq 0$ in $\mathscr{O}$ was made only to give a physical meaning to the propagation process.

The conditions $m \leq 5$ and $x \in L^\infty$ might seem unnatural, but they are technical assumptions required by the work [15] on which the present proof essentially relies.

3.9.1 Proof of Theorem 3.9.1

Without loss of generality we may take $\xi_0 = 0 \in \mathscr{O}$ and set $B_r = B_r(0)$. The method of the proof relies on some sharp integral energy type estimates of $X = X(t)$ on arbitrary balls $B_r \subset \mathscr{O}$.

It is convenient to rewrite, as in previous situation, Eq. (3.93) as a deterministic equation with random coefficients. To this aim we consider the transformation (3.45)

$$y(t) = e^{W(t)} X(t), \ t \geq 0, \tag{3.100}$$

where $W(t) = -\sum_{k=1}^{N} \mu_k e_k \beta_k(t)$.

Then we have, see (3.46),

$$\begin{cases} \dfrac{dy}{dt} - e^W \Delta(y^m e^{-mW}) + \dfrac{1}{2} \mu y = 0, \ t > 0, \ \mathbb{P}\text{-a.s.,} \\ y(0) = x, \\ y^m \in H_0^1(\mathscr{O}), \ \forall t > 0, \ \mathbb{P}\text{-a.s.,} \end{cases} \tag{3.101}$$

where

$$\mu = \sum_{k=1}^{N} \mu_k^2 e_k^2. \tag{3.102}$$

By Theorem 3.6.1, we have $\mathbb{P}$-a.s.

$$y \geq 0, \ y^m(t)e^{-mW(t)} \in H_0^1 \cap L^{\frac{m+1}{m}}, \text{ a.e. } t \geq 0. \tag{3.103}$$

As a matter of fact, one has

Lemma 3.9.2 *Assume that* $1 \leq d \leq 3$ *and* $m \in [1, 5]$. *Then, if* $x \in L^\infty$, *the solution* y *to* (3.101) *satisfies* $\mathbb{P}$-*a.s. for every* $T > 0$

$$y \in L^\infty((0, T) \times \mathcal{O}) \cap C([0, T]; H^{-1}), \tag{3.104}$$

$$y^m \in L^2(0, T; H_0^1), \ \frac{dy}{dt} \in L^2(0, T; H^{-1}). \tag{3.105}$$

Moreover, for every $T \in (0, 1]$, $\alpha \in (0, \frac{1}{2})$, $R > 0$, *there exist constants* $c_1, c_2 > 0$ *depending on* α, R, $\mathcal{O}$, $|x|_\infty$, $\max_{1 \leq k \leq N}(|e_k|_\infty, |\nabla e_k|_\infty, |\Delta e_k|_\infty)$, *but not on* T *such that* $\mathbb{P}$-*a.s. on* $\Omega_{H,R}^\alpha$,

$$\|y\|_{L^\infty((0,T)\times\mathcal{O})} \leq c_1 \exp\left[c_2 \max_{1 \leq k \leq N} \sup_{t \in [0,T]} |\beta_k(t)| \right]. \tag{3.106}$$

The first part of Lemma 3.9.2 is just Theorem 3.6.1, while (3.101) follows by Theorem 3.6.2.

Before we introduce our crucial energy functional ϕ in (3.113) below and explaining the idea of the proof subsequently, we need some preparations by a few estimates on the solution y to (3.101). Everywhere in the following we fix $\alpha \in (0, \frac{1}{2})$, $\alpha > 0$ and assume that $x \geq 0$ so that (3.103) holds and fix $T \in (0, 1]$.

By Green's formula, it follows from (3.101) that

$$\frac{1}{m+1} \int_{\mathcal{O}} y^{m+1}(t, \xi) \psi(\xi) d\xi + \int_0^t ds \int_{\mathcal{O}} \nabla(y^m e^{-mW}) \cdot \nabla(e^W y^m \psi) d\xi$$

$$+ \frac{1}{2} \int_0^t ds \int_{\mathcal{O}} \mu y^{m+1} \psi \, d\xi$$

$$= \frac{1}{m+1} \int_{\mathcal{O}} x^{m+1}(\xi) \psi(\xi) d\xi, \ t \in (0, T), \tag{3.107}$$

for all $\psi \in C_0^\infty(\mathcal{O})$.

Fix $r > 0$ and let $\rho_\varepsilon \in C^\infty(R^+)$ be a cut-off function such that $\rho_\varepsilon(s) = 1$ for $0 \le s \le r + \varepsilon$, $\rho_\varepsilon(s) = 0$ for $s \ge r + 2\varepsilon$ and for $\chi_\varepsilon = \mathbb{1}_{[r+\varepsilon, r+2\varepsilon]}$,

$$\lim_{\varepsilon \to 0} \left| \rho'_\varepsilon(s) + \frac{1}{\varepsilon} \right| \chi_\varepsilon(s) = 0, \tag{3.108}$$

uniformly in $s \in [0, \infty)$. Roughly speaking, this means that ρ_ε is a smooth approximation of the function $\gamma_\varepsilon(s) = 1$ on $[0, r + \varepsilon]$, $\gamma_\varepsilon(s) = 0$ on $[r + 2\varepsilon, \infty)$, $\gamma_\varepsilon(s) = -\frac{1}{\varepsilon}(s - r - \varepsilon) + 1$ on $[r + \varepsilon, r + 2\varepsilon]$.

If in (3.107) we take $\psi = \rho_\varepsilon(|\xi|)$ (for ε small enough), setting $\psi_\varepsilon(\xi) = \rho_\varepsilon(|\xi|)$, $\xi \in \mathcal{O}$, we obtain that

$$\frac{1}{m+1} \int_\mathcal{O} (y(t, \xi))^{m+1} \rho_\varepsilon(|\xi|) d\xi + \int_0^t ds \int_\mathcal{O} \nabla(ye^{-W})^m \cdot \nabla(e^W y^m \psi_\varepsilon) d\xi$$

$$+ \frac{1}{2} \int_0^t ds \int_\mathcal{O} Wy^{m+1} \psi_\varepsilon d\xi = \frac{1}{m+1} \int_\mathcal{O} x^{m+1} \psi_\varepsilon d\xi. \tag{3.109}$$

On the other hand, we have

$$\int_\mathcal{O} \nabla(ye^{-W})^m \cdot \nabla(e^W y^m \psi_\varepsilon) d\xi = \int_\mathcal{O} |\nabla(ye^{-W})^m|^2 \psi_\varepsilon e^{(m+1)W} d\xi$$

$$+ (m+1)\frac{1}{2} \int_\mathcal{O} (\nabla(ye^{-W})^m \cdot \nabla W) e^W y^m \psi_\varepsilon d\xi$$

$$+ \int_\mathcal{O} (\nabla(ye^{-W})^m \cdot v)(s, \xi) \rho'_\varepsilon(|\xi|)(e^W y^m)(s, \xi) d\xi, \tag{3.110}$$

where $v(\xi) = \frac{\xi}{|\xi|}$. (Since $\mu \in C^2(\overline{\mathcal{O}})$, the above calculation is justified.)

Everywhere in the following, the estimates are taken $\mathbb{P}$-a.s. on the set $\Omega_{H,R}^\alpha \cap \Omega_T^{\delta(R)}$.

We set $B_r^\varepsilon = B_{r+2\varepsilon} \setminus B_{r+\varepsilon}$. Then, by (3.109), (3.110), we see that

$$\frac{1}{m+1} \int_{B_{r+\varepsilon}} y^{m+1}(t, \xi) d\xi + \int_0^t ds \int_{B_{r+2\varepsilon}} \psi_\varepsilon e^{(m+1)W} |\nabla(ye^{-W})^m|^2 d\xi \, ds$$

$$+ \frac{1}{2} \int_0^t ds \int_{B_{r+2\varepsilon}} \psi_\varepsilon Wy^{m+1} d\xi \, ds$$

$$\le \frac{1}{m+1} \int_{B_{r+2\varepsilon}} \psi_\varepsilon x^{m+1} d\xi \tag{3.111}$$

$$- (m+1) \int_0^t \int_{B_{r+2\varepsilon}} (\nabla(ye^{-W})^m \cdot W) \psi_\varepsilon e^W y^m d\xi \, ds$$

$$- \int_0^t \int_{B_r^\varepsilon} (\nabla(ye^{-W})^m \cdot v)(s, \xi)(e^\mu y^m)(s, \xi) \rho'_\varepsilon(|\xi|) d\xi \, ds.$$

On the other hand, we have

$$\int_0^t \int_{B_r^\varepsilon} |(\nabla(ye^{-W})^m \cdot v)e^W y^m \rho_\varepsilon'(| \cdot |)|d\xi \, ds$$

$$\leq \left(\int_0^t \int_{B_r^\varepsilon} |\rho_\varepsilon'(| \cdot |)| \, |\nabla(ye^{-W})^m|^2 e^{(m+1)W} d\xi \, ds \right)^{\frac{1}{2}} \tag{3.112}$$

$$\times \left(\int_0^t \int_{B_r^\varepsilon} e^{(1-m)W} y^{2m} |\rho_\varepsilon'(| \cdot |)|d\xi \, ds \right)^{\frac{1}{2}}.$$

We introduce the energy function

$$\phi(t, r) = \int_0^t \int_{B_r} |\nabla(ye^{-W})^m|^2 e^{(m+1)W} d\xi \, ds, \ t \in [0, T], \ r \geq 0. \tag{3.113}$$

In order to prove (3.99), our aim in the following is to show that ϕ satisfies a differential inequality of the form

$$\frac{\partial \phi}{\partial r}(t, r) \geq Ct^{\theta-1}(\phi(t, r))^\delta \text{ on } \Omega_{H,R}^\alpha \cap \Omega_T^{\delta(R)} \text{ for } t \in [0, T], \ r \in [0, r_0],$$

where $0 < \theta < 1$ and $0 < \delta < 1$ and from which (3.99) will follow.

Taking into account that function ϕ is absolutely continuous in r, we have by (3.108), a.e. on $(0, r_0)$,

$$\lim_{\varepsilon \to 0} \int_0^t \int_{B_r^\varepsilon} |\rho_\varepsilon'(| \cdot |)| |\nabla(ye^{-W})^m|^2 e^{(m+1)W} d\xi \, ds = \frac{\partial \phi}{\partial r}(t, r).$$

Then, letting $\varepsilon \to 0$ in (3.111), (3.112), we obtain that

$$\frac{1}{m+1} \int_{B_r} y^{m+1}(t, \xi)d\xi + \phi(t, r) + \frac{1}{2} \int_0^t \int_{B_r} Wy^{m+1} d\xi \, ds$$

$$\leq \frac{1}{m+1} \int_{B_r} x^{m+1} d\xi - (m+1) \int_0^t \int_{B_r} (\nabla(ye^{-W})^m \cdot \nabla W)e^W y^m d\xi \, ds \tag{3.114}$$

$$+ \left(\frac{\partial \phi}{\partial r}(t, r) \right)^{\frac{1}{2}} \left(\int_0^t ds \int_{\Sigma_r} y^{2m} e^{(1-m)W} d\xi \right)^{\frac{1}{2}},$$

on $\Omega_{H,R}^\alpha \cap \Omega_T^{\delta(R)}$, $t \in [0, T]$, $r \in [0, r_0]$.

In order to estimate the right-hand side of (3.114), we introduce the following notations

$$K(t,r) = \frac{1}{2} \int_0^t \int_{B_r} W y'^{m+1} ds\, d\xi \tag{3.115}$$

$$H(t,r) = \sup \left\{ \frac{1}{m+1} \int_{B_r} y^{m+1}(s,\xi) d\xi,\ 0 \le s \le t \right\}, \tag{3.116}$$

and note that by assumption (3.96) we have

$$K(t,r) \ge \frac{1}{2} \rho \int_0^t \int_{B_r} \dot{y}^{m+1} d\xi\, ds,\ \forall t \in [0,T],\ r \in [0,r_0]. \tag{3.117}$$

Then (3.114) yields, for $r \in (0,r_0]$,

$$H(t,r) + \phi(t,r) + K(t,r) \le (m+1) \int_0^t \int_{B_r} |(\nabla(ye^{-W})^m \cdot \nabla W)e^W y^m| d\xi\, ds$$
$$+ \left(\frac{\partial \phi}{\partial r}(t,r)\right)^{\frac{1}{2}} \left(\int_0^t \int_{\Sigma_r} y^{2m} e^{(1-m)W} d\xi\, ds\right)^{\frac{1}{2}} \tag{3.118}$$

because $x \equiv 0$ on B_r. We note that, by the trace theorem, the surface integral arising in the right-hand side of formula (3.118) is well defined because $\nabla(ye^{-W})^m \in L^2([0,T] \times \mathcal{O})$ and, by Lemma 3.9.2, $y \in L^\infty((0,T) \times \mathcal{O})$ $\mathbb{P}$-a.s.

Now, we are going to estimate the right-hand side of (3.118).

By Cauchy–Schwarz and (3.117), we have

$$\int_0^t \int_{B_r} |(\nabla(ye^{-W})^m \cdot \nabla W)e^W y^m| d\xi\, ds$$
$$\le \|y^{m-1} e^{(1-m)W} |\nabla W|^2\|_{L^\infty((0,T)\times\mathcal{O})}^{1/2}$$
$$\times \left(\int_0^t ds \int_{B_r} |\nabla(y^m e^{-mW})|^2 e^{(m+1)W} d\xi\right)^{\frac{1}{2}} \left(\int_0^t ds \int_{B_r} y^{m+1} d\xi\right)^{\frac{1}{2}}$$
$$\le (2\rho^{-1})^{1/2} \|y^{m-1} e^{(1-m)W} |\nabla W|^2\|_{L^\infty((0,T)\times\mathcal{O})}^{1/2} (\phi(t,r))^{\frac{1}{2}} (K(t,r))^{\frac{1}{2}}$$
$$\le \frac{1}{2(m+1)} (\phi(t,r) + K(t,r)),\ \forall t \in (0,T],\ r \in (0,r_0],\ \text{on } \Omega_{H,R}^\alpha \cap \Omega_T^{\delta(R)}, \tag{3.119}$$

by the definition of $\delta(R)$.

By (3.118), it follows that

$$H(t,r) + \phi(t,r) + K(t,r)$$

$$\le \left(\frac{\partial \phi}{\partial r}(t,r)\right)^{\frac{1}{2}} \left(\int_0^t ds \int_{\Sigma_r} y^{2m} e^{(1-m)W} d\xi\right)^{\frac{1}{2}} \tag{3.120}$$
$$\forall t \in [0,T],\ r \in [0,r_0],\ \text{on } \Omega_{H,R}^\alpha \cap \Omega_T^{\delta(R)}.$$

In order to estimate the surface integral from the right-hand side of (3.120), we invoke the following interpolation-trace inequality (see, e.g., Lemma 2.2 in [54])

$$|z|_{L^2(\Sigma_r)} \le C(|\nabla z|_{L^2(B_r)} + |z|_{L^{\sigma+1}(B_r)})^\theta |z|_{L^{\sigma+1}(B_r)}^{1-\theta}, \tag{3.121}$$

for all $\sigma \in [0,1]$ and $\theta = (d(1-\sigma) + \sigma + 1)/(d(1-\sigma) + 2(\sigma+1))$. Clearly, $\theta \in [\frac{1}{2}, 1)$.

We shall apply this inequality for $z = (y^m e^{-W})^m$ and $\sigma = \frac{1}{m}$. We obtain, by (3.23) that

$$\left(\int_{\Sigma_r} y^{2m} e^{(1-m)W} d\xi \right)^{\frac{1}{2}} \le \|e^{(1+m)W}\|_{L^\infty((0,T)\times\mathcal{O})}^{1/2} \left(\int_{\Sigma_r} (ye^{-W})^{2m} d\xi \right)^{\frac{1}{2}}$$

$$\le C\|e^{(1+m)W}\|_{L^\infty((0,T)\times\mathcal{O})}^{1/2} (|\nabla(ye^{-W})^m|_{L^2(B_r)} + |y^m e^{-mW}|_{L^{\frac{m+1}{m}}(B_r)})^\theta$$

$$\times |y^m e^{-mW}|_{L^{\frac{m+1}{m}}(B_r)}^{1-\theta}$$

$$\le \widetilde{C} \left(\left(\int_{B_r} |\nabla(y^m e^{-mW})|^2 e^{(m+1)W} d\xi \right)^{\frac{1}{2}} + H^{\frac{m}{m+1}}(t,r) \right)^\theta$$

$$(H^{\frac{m}{m+1}}(t,r))^{1-\theta}, \quad \text{on } \Omega_{H,R}^\alpha \cap \Omega_T^{\delta(R)},$$

where, as will be the case below, $\widetilde{C}$ is a positive function of $\omega \in \Omega_{H,R}^\alpha \cap \Omega_T^{\delta(R)}$, independent of t and r, which may change below from line to line.

Integrating over $(0,t)$ and applying first Minkowski's (since $\theta \ge \frac{1}{2}$) and then Hölder's inequality yields

$$\left(\int_0^t ds \int_{\Sigma_r} y^{2m} e^{(1-m)W} d\xi \right)^{\frac{1}{2}}$$

$$\le \widetilde{C} \left(\int_0^t ds \left(\int_{B_r} |\nabla(y^m e^{-mW})|^2 e^{(m+1)W} d\xi + H^{\frac{2m}{m+1}}(s,r) \right)^\theta H^{\frac{2m(1-\theta)}{m+1}}(s,r) \right)^{\frac{1}{2}}$$

$$\le \widetilde{C} H^{\frac{m(1-\theta)}{m+1}}(t,r) t^{\frac{1-\theta}{2}} ((\phi(t,r))^{\frac{1}{2}} + H^{\frac{m}{m+1}}(t,r))^\theta, \quad \text{on } \Omega_{H,R}^\alpha \cap \Omega_T^{\delta(R)}.$$

Substituting the latter into (3.120), we obtain that

$$\phi + H \le \widetilde{C} t^{\frac{1-\theta}{2}} \left(\frac{\partial\phi}{\partial r} \right)^{\frac{1}{2}} (\phi^{\frac{1}{2}} + H^{\frac{m}{m+1}})^\theta H^{\frac{m(1-\theta)}{m+1}}$$

$$\le \widetilde{C} t^{\frac{1-\theta}{2}} \left(\frac{\partial\phi}{\partial r} \right)^{\frac{1}{2}} \left(\phi^{\frac{1}{2}} H^{\frac{m(1-\theta)}{(m+1)\theta}} + H^{\frac{m}{(m+1)\theta}} \right)^\theta, \tag{3.122}$$

$$\forall t \in [0,T], \ r \in [0, r_0], \ \text{on } \Omega_{H,R}^\alpha \cap \Omega_T^{\delta(R)}.$$

On the other hand, for $H_0 = H(T, r_0)$, we have the estimate

$$\phi^{\frac{1}{2}} H^{\frac{m(1-\theta)}{(m+1)\theta}} + H^{\frac{m}{(m+1)\theta}} \leq \phi^{\frac{1}{2}} H^{\frac{m(1-\theta)}{(m+1)\theta}} + H_0^{\frac{m}{m+1}-\frac{1}{2}} H^{\frac{m(1-\theta)}{(m+1)\theta}+\frac{1}{2}} \leq \widetilde{C}(\phi + H)^{\frac{1}{2}+\frac{m(1-\theta)}{(m+1)\theta}},$$

where $\widetilde{C} := 2 \max(1, H_0^{\frac{m-1}{2(m+1)}})$ and where we used that by Young's inequality, for all $p, q \in (0, \infty)$,

$$\phi^p H^q \leq (\phi + H)^{p+q}.$$

Substituting the latter into (3.122) yields

$$\phi + H \leq \widetilde{C} t^{\frac{1-\theta}{2}} \left(\frac{\partial \phi}{\partial r}\right)^{\frac{1}{2}} (\phi + H)^{\frac{\theta}{2}+\frac{m(1-\theta)}{m+1}} \quad \text{on } (0, T) \times (0, r_0) \times \Omega_{H,R}^{\alpha} \cap \Omega_T^{\delta(R)},$$

and therefore

$$\left(\frac{\partial \phi}{\partial r}(t, r)\right)^{\frac{1}{2}} \geq \widetilde{C} t^{\frac{\theta-1}{2}} (\phi(t, r))^{\frac{2-\theta}{2}-\frac{m(1-\theta)}{m+1}} \quad \text{on } (0, T) \times (0, r_0) \times \Omega_{H,R}^{\alpha} \cap \Omega_T^{\delta(R)}. \tag{3.123}$$

Equivalently,

$$\frac{\partial \varphi}{\partial r}(t, r) \geq \widetilde{C} t^{\theta-1}, \quad \text{on } (0, T) \times (r(t), r_0) \times \Omega_{H,R}^{\alpha} \cap \Omega_T^{\delta(R)}, \tag{3.124}$$

where

$$\varphi(t, r) = (\phi(t, r))^{\theta+\frac{2m(1-\theta)}{m+1}-1}, \tag{3.125}$$

and

$$r(t) := \inf\{r \geq 0 \mid \phi(t, r) > 0\} \wedge r_0.$$

We note that, by continuity,

$$\phi(t, r(t)) = 0$$

and that, since $\phi(t, r)$ is increasing in t and r, we have $\phi(t, r) > 0$, if $r > r(t)$, and that $t \mapsto r(t)$ is decreasing in t. Furthermore, the same is true for φ defined in (3.125), since $\theta + \frac{2m(1-\theta)}{m+1} - 1 > 0$, because $0 < \theta < 1$ and $m > 1$.

Moreover, by (3.23) and (3.122) we see that

$$X(t, \xi) = 0 \qquad \text{for} \quad \xi \in B_{r(t)}.$$

We recall that $r(t) = r(t, \omega)$ depends on $\omega \in \Omega$. Now, fix $\omega \in \Omega_{H,R}^\alpha \times \Omega_T^{\delta(R)}$. Our aim is to show that

$$\exists \, t(\omega) \in (0, T] \text{ such that } r(t, \omega) > 0, \ \forall t \in [0, t(\omega)]. \qquad (3.126)$$

Since we already noted that $\phi(t, r) > 0$, if $r > r(t)$, by (3.101), (3.125) and (3.100), we deduce the property in (3.99) from (3.126). To show (3.36), we first note that by (3.124) for all $t \in (0, T)$

$$\varphi(t, r_0)(\omega) \geq \widetilde{C} t^{\theta-1}(r_0 - r(t, \omega)),$$

hence

$$r(t, \omega) \geq r_0 - (\widetilde{C}(\omega))^{-1} t^{1-\theta} \varphi(t, r_0)(\omega).$$

So, because $0 < \theta < 1$, we can find $t = t(\omega) \in (0, T)$, small enough, so that the right-hand side is strictly positive. Now, (3.36) follows, since, as noted earlier, $t \mapsto r(t, \omega)$ is decreasing in t, which completes the proof of (3.99). By elementary considerations for $\delta > 0$, we have

$$\mathbb{P}(\Omega_T^\delta) \geq 2^N \left(1 - \sqrt{\frac{T}{2\pi\delta^2}} \, e^{-\delta^2/(2T)} \right)^N.$$

Hence $\Omega_T^\delta \nearrow \Omega$ as $T \to 0$ up to a $\mathbb{P}$-zero set and the last part of the assertion also follows.

Remark 3.9.3 In the deterministic case, for $\mathcal{O} = \mathbb{R}^d$ the finite speed propagation property: support $\{x\} \subset B_{r_0}(\xi_0) \implies$ support $\{X(t)\} \subset B_{r(t)}(\widetilde{\xi}_0)$ for some $\widetilde{\xi}_0 \in \mathbb{R}^d$ and $r = r(t)$, follows by the comparison principle $X(t, \xi) \leq U(t + \tau, \xi - \widetilde{\xi}_0)$, where $U = U(t, \xi)$ is the Barenblatt source solution

$$U(t, \xi) = t^{-\frac{d}{(m-1)d+2}} \left[C - \frac{m-1}{2m((m-1)d+2)} \frac{|\xi|^2}{t^{\frac{2}{(m-1)d+2}}} \right]_+^{\frac{1}{m-1}} \qquad (3.127)$$

(see [89]) and which has the support in $\{(t, \xi); \ |\xi|^2 \leq C_1 t^{\frac{2}{(m-1)d+2}}\}$.

At least in the simpler case, where the noise is not function valued, i.e. independent of ξ, this is similar in the stochastic case. More precisely, for $m = 2$, $d = 1$, $\mathcal{O} = \mathbb{R}^1$ and $W(t) = \beta(t)$ =standard, real-valued Brownian motion, the function

$$Z(t, \xi) = U \left(\int_0^t k(s)ds, \xi \right) k(t), \ k(t) = e^{\beta(t) - \frac{1}{2}t}$$

is a solution to (3.93) and support $Z \subset \left\{ (t, \xi); |\xi|^2 \leq C_1 \left(\int_0^t k(s)ds \right)^{\frac{2}{3}} \right\}$ (see [78] for details). However, on bounded domains, it is not clear, whether this is applicable.
□

Remark 3.9.4 We refer to [3, 54, 89, 90] for corresponding localization results in deterministic case. As a matter of fact the energy method used here was introduced by S. N. Antonsev and developed in [3].

The finite dimensional structure of the Wiener process $W(t)$ was essential for the present approach, which is based on sharp estimates on solutions to Eq. (3.101). A direct application of the above energy method in $L^2(\Omega; L^2(0, T; H^{-1}))$ failed for general cylindrical Wiener processes $W(t)$.
□

3.10 The Logarithmic Diffusion Equation

We consider here the nonlinear diffusion equation with linear multiplicative noise (logarithmic diffusion equation)

$$\begin{cases} dX(t) = \Delta \log(X(t))dt + \sigma(X(t))dW(t), & \text{in } (0, T) \times \mathcal{O} \\[2mm] X(0) = x \in H^{-1}, \\[2mm] X(t) = 1, & \text{on } (0, T) \times \partial\mathcal{O} \end{cases} \tag{3.128}$$

where

$$\begin{aligned} \sigma(X(t))dW(t) &= KX(t)(-A)^{-\gamma}dW(t) \\[2mm] &= K\sum_{h=1}^{\infty} \alpha_h^{-\gamma}(X(t)e_h)dW_h(t). \end{aligned} \tag{3.129}$$

Here K is a positive constant, $\{\alpha_h\}$ are the eigenvalues of $A = -\Delta$ with Dirichlet boundary conditions and $\gamma > \frac{3}{d}$ (see Example 2.1.2). Moreover W is a cylindrical Wiener process in H^{-1}.

Definition 3.10.1 The process X is called a *distributional solution* of (3.128) if the following conditions hold.

(i) $X \in L^2_W(\Omega; C([0, T]; H^{-1} \cap L^1((0, T) \times \mathcal{O} \times \Omega))$.
(ii) $X > 0$ a.e. in $(0, T) \times \mathcal{O} \times \Omega$.
(iii) $\log X \in L^2_W(0, T; L^2(\Omega; H_0^1))$.

(iv) $\int_0^t \log X(s)ds \in L^2_W(\Omega; C([0,T]; H^1_0)).$

(v) We have

$$X(t) = x + \Delta \int_0^t \log(X(s))ds + K \int_0^t (X(s))A^{-\gamma}dW(s), \ \forall\, t \in [0,T], \ \mathbb{P}\text{-a.s.,}$$

(3.130)

where Δ is considered in sense of distributions on $\mathcal{O}$.

As mentioned earlier in Sect. 1.1, Eq. (3.128) can be viewed as a superfast diffusion equation as the limit case $m = 0$ of Eq. (1.5) that is

$$dX(t) = \text{div}\left(\frac{\nabla X(t)}{X(t)}\right) + KX(t)A^{-\gamma}dW(t)$$

(3.131)

and it models the dynamics of plasma in a magnetic field perturbed by a multiplicative Gaussian noise. We also recall that the deterministic version of (3.130) (equivalently (3.128)) arises in the Riemannian geometry, as a model for the evolution of conformally flat metric driven by its Ricci curvature flow. Taking into account that by the rescaling transformation $X = e^{KY(t)A^{-\gamma}W(t)}$, Eq. (3.128) reduces formally to the deterministic equation

$$\frac{\partial Y}{\partial t} = e^{-W}\Delta(\log Y) - e^{-W}\Delta W + \frac{K^2}{2}\sum_{h=1}^{\infty}\alpha_h^{-2\gamma}e_h^2 Y, \quad \forall\, \omega \in \Omega.$$

one might obtain a similar geometric interpretation to stochastic equation (3.128).

Here is the main result of the section.

Theorem 3.10.2 *Let* $x \in L^2$ *such that* $x > 0$, $x \log x \in L^2$. *Then for each* $T > 0$ *there is a unique strong solution* X *to* (3.128). *Moreover* $X|\log X| \in L^1((0,T) \times \Omega \times \mathcal{O})$

Proof Arguing as in Sect. 3.3 we consider for any $\epsilon > 0$ the approximating problem

$$\begin{cases} dX_\epsilon(t) = \Delta(\beta_\epsilon(X_\epsilon(t)) + \epsilon X_\epsilon(t))dt + \sigma(X_\epsilon(t))\,dW(t), \\[2mm] X_\epsilon(0) = x, \end{cases}$$

(3.132)

where β_ϵ are the Yosida approximation of the maximal monotone function $\beta : \mathbb{R} \to \mathbb{R}$

$$\beta(r) = \begin{cases} \log r & \text{if } r > 0 \\ \varnothing & \text{if } r \le 0. \end{cases}$$

(3.133)

By Theorem 2.5.1 we know that Eq. (3.132) has a unique strong solution $X_\epsilon \in L_W^2(\Omega; C([0,T]; H^{-1})) \cap L^2(0,T; L^2(\Omega; L^2))$, such that $\beta_\epsilon(X_\epsilon) + \epsilon X_\epsilon \in L_W^2(0,T; L^2(\Omega; H_0^1))$.

We are going to prove that for $\epsilon \to 0$, $\{X_\epsilon\}$ is convergent in a suitable norm to a strong solution X to (3.128). To this end we need a few a priori estimates. Estimates for $\|X_\epsilon(t)\|_{H^{-1}}$ is provided by Lemma 2.4.2. In particular we recall that

$$\frac{1}{2}\|X_\epsilon(t)\|_{-1}^2 + \int_0^t \int_{\mathcal{O}} (\beta_\epsilon X_\epsilon(s) + \epsilon X_\epsilon(s)) X_\epsilon(s) \, d\xi \, ds$$

$$= \frac{1}{2}\|x\|_{-1}^2 + \frac{K}{2} \sum_{h=1}^{\infty} \alpha_k^{-\gamma} \int_0^t \int_{\mathcal{O}} \langle X_\epsilon(s), X_\epsilon(s) e_k \rangle_{-1} dW_k(s) \, d\xi \qquad (3.134)$$

$$+ \frac{K^2}{2} \sum_{k=1}^{\infty} \alpha_k^{-2\gamma} \int_0^t \int_{\mathcal{O}} |X_\epsilon(s) e_k|_{-1}^2 d\xi \, ds.$$

Moreover, we need an additional estimate for $\varphi_\epsilon(X_\epsilon)$ where

$$\varphi_\epsilon(x) = \int_{\mathcal{O}} j_\epsilon(x) d\xi,$$

and

$$j_\epsilon(r) = \int_0^r \beta_\epsilon(s) ds + \frac{\epsilon}{2} r^2, \quad \forall r \in \mathbb{R}.$$

Lemma 3.10.3 *The following identity holds*

$$\beta_\epsilon'(r) = \frac{1}{(1+\epsilon\beta)^{-1}(r) + \epsilon}, \quad \forall r \in \mathbb{R}, \ \epsilon > 0. \qquad (3.135)$$

Proof Write

$$\beta_\epsilon(r) = \beta((1+\epsilon\beta)^{-1}(r)) = \log((1+\epsilon\beta)^{-1}(r)). \qquad (3.136)$$

Then

$$\beta_\epsilon'(r) = \frac{D(1+\epsilon\beta)^{-1}(r)}{(1+\epsilon\beta)^{-1}(r)}, \quad \forall r \in \mathbb{R}, \ \epsilon > 0. \qquad (3.137)$$

To compute $D(1+\epsilon\beta)^{-1}(r)$ we set $s(r) := (1+\epsilon\beta)^{-1}(r)$, $r \in \mathbb{R}$, so that

$$s(r) + \epsilon\beta(s(r)) = r,$$

which, differentiating with respect to r, yields

$$s'(r) + \epsilon \frac{s'(r)}{s(r)} = 1.$$

So,

$$D(1 + \epsilon\beta)^{-1}(r) = s'(r) = \frac{s(r)}{s(r) + \epsilon}$$

Finally, substituting this into (3.137), we arrive at the conclusion. $\square$

Now by Itô's formula we have

$$\mathbb{E}\,\varphi_\epsilon(X_\epsilon(t)) + \mathbb{E}\int_0^t \int_{\mathscr{O}} |\nabla\beta_\epsilon(X_\epsilon(s)) + \epsilon\nabla X_\epsilon(s)|^2 d\xi\, ds \qquad (3.138)$$

$$= \varphi_\epsilon(x) + \frac{K^2}{2}\sum_{k=1}^{\infty}(\alpha_k)^{-2\gamma}\,\mathbb{E}\int_0^t \int_{\mathscr{O}} |X_\epsilon(s)e_k|^2(\beta'_\epsilon(X_\epsilon(s)) + \epsilon)d\xi\, ds.$$

(As in previous cases, in order to be rigorous to get (3.138) we should proceed into two steps. The first is for φ_ϵ replaced by $x \mapsto \varphi_\epsilon((1 - \nu\Delta)^{-1}x)$ and in the second one we pass to the limit $\nu \to 0$.)

Now, taking into account (3.138), yields

$$\mathbb{E}\int_{\mathscr{O}} j_\epsilon(X_\epsilon(t))d\xi + \mathbb{E}\int_0^t \int_{\mathscr{O}} (|\nabla\beta_\epsilon(X_\epsilon(s))|^2 + |\epsilon\nabla X_\epsilon(s)|^2)ds\, d\xi \qquad (3.139)$$

$$\leq \int_{\mathscr{O}} j(x)d\xi + \frac{\epsilon}{2}\int_{\mathscr{O}} x^2(\xi)d\xi + C\int_0^t \int_{\mathscr{O}} X_\epsilon(s)g_\epsilon(X_\epsilon(s))ds\, d\xi,$$

where $j(x) = x\log x - 1$, $C > 0$ is independent of ϵ and

$$g_\epsilon(r) = \frac{(1 + \epsilon\beta)^{-1}r}{(1 + \epsilon\beta)^{-1}(r) + \epsilon} + \epsilon, \quad \forall\, r \in \mathbb{R},\ \epsilon > 0.$$

It is easily seen that $g'_\epsilon(r) > 0$ for all > 0 and that

$$\lim_{r\to\infty} g_\epsilon(r) = \lim_{r\to\infty} \frac{(1 + \epsilon\beta)^{-1}(r)}{(1 + \epsilon\beta)^{-1}(r) + \epsilon} + \epsilon = \frac{1}{1 + \epsilon} + \epsilon.$$

Hence $g_\epsilon(X_\epsilon(s)) \leq 1 + \epsilon^2$ and so, (3.139) yields

$$\mathbb{E}\int_{\mathscr{O}} j_\epsilon(X_\epsilon(t))d\xi + \mathbb{E}\int_0^t \int_{\mathscr{O}} |\nabla\beta_\epsilon(X_\epsilon(s))|^2 ds\, d\xi$$

$$\leq C(1 + \epsilon^2) + C\int_{\mathscr{O}}(j(x) + \epsilon x^2)d\xi + C\int_0^t \int_{\mathscr{O}} |X_\epsilon(s)|ds\, d\xi \qquad (3.140)$$

$$\leq C_T, \quad \forall\, t \in [0, T].$$

Now we can proceed as in Sect. 2.4 (Proposition 2.4.4 and Theorem 2.5.1) to prove that

$$\mathbb{E} \sup_{0 \le t \le T} \|X_\epsilon(t) - X_{\epsilon'}(t)\|_{-1} = \delta(\epsilon, \epsilon') \to 0 \text{ as } \epsilon, \epsilon' \to 0. \tag{3.141}$$

On the other hand, recalling that (see (1.37))

$$j_\epsilon(X_\epsilon) = \frac{\epsilon}{2}|X_\epsilon - (1 + \epsilon\beta)^{-1}X_\epsilon|^2 + j((1 + \epsilon\beta)^{-1}X_\epsilon),$$

we get by (3.140) that

$$\mathbb{E} \int_{\mathcal{O}} ((1 + \epsilon\beta)^{-1}X_\epsilon)(\log(1 + \epsilon\beta)^{-1}X_\epsilon) - 1)d\xi \le C. \tag{3.142}$$

In particular, it follows by (3.142) via Dunford–Pettis compactness criterium in $L^1((0, T) \times \Omega \times \mathcal{O})$ (Theorem 1.2.12) that $\{(1 + \epsilon\beta)^{-1}X_\epsilon\}$ is weakly compact in $L^1((0, T) \times \Omega \times \mathcal{O})$. Indeed, since $\lim_{r \to \infty} \frac{j(r)}{r} = \infty$ it follows that the set $\{(1 + \epsilon\beta)^{-1}X_\epsilon\}_{\epsilon > 0}$ is bounded and equi-integrable.

Now by estimates (3.138), (3.140), (3.143) it follows that there are

$$X \in L^2_W(\Omega; C([0, T]; H^{-1}) \cap L^1((0, T) \times \Omega \times \mathcal{O})$$

$$\eta \in L^2_W(0, T; L^2(\Omega; H^1_0)),$$

such that for a subsequence $\epsilon \to 0$

$$\begin{cases} X_\epsilon \to X \quad \text{strongly in } L^2(\Omega; C([0, T]; H^{-1})) \\[2mm] (1 + \epsilon\beta)^{-1}X_\epsilon \to X \quad \text{weakly in } L^1((0, T) \times \Omega \times \mathcal{O}) \\[2mm] \beta_\epsilon(X_\epsilon) + \epsilon X_\epsilon \to \eta \quad \text{weakly in } L^2(0, T; L^2(\Omega; H^1_0)) \\[2mm] (1 + \epsilon\beta)^{-1}(X_\epsilon) - X_\epsilon \to 0 \quad \text{strongly in } L^2(0, T; L^2(\Omega; L^2)). \end{cases} \tag{3.143}$$

Moreover, letting $\epsilon \to 0$ into (3.132) we see that

$$X(t) = x + \Delta \int_0^t \eta(s)ds + k \int_0^t (X(s))A^{-\gamma}dW(s), \ \forall t \in [0, T], \ \mathbb{P}\text{-a.s..} \tag{3.144}$$

To conclude the proof of existence it remains to be shown that

$$\eta = \log X \quad \text{a.e. in } \Omega \times (0, T) \times \mathcal{O}. \tag{3.145}$$

Since the realization of operator $x \rightarrow -\Delta\beta(x)$ is maximal monotone in $L^2((0, T) \times \Omega \times H^{-1})$ to get (3.145) it suffices to take into account (3.143) and that it is strongly-weakly closed in

$$L^2((0, T) \times \Omega \times H_0^1) \times L^2((0, T) \times \Omega \times H^{-1}).$$

As regards uniqueness, it follows from a straightforward argument see e.g. Theorem 3.2.2. □

3.11 Comments and Bibliographical Remarks

The main existence results, Theorem 3.4.1, Theorem 3.5.5 were proved first in a related form in [21, 23]. Other results were established also in [44, 47, 53, 62, 68, 84].

The finite extinction of solutions to fast diffusion equations, Theorem 3.7.3, was established in [23, 25], and for the self-organized criticality equation in [24]. The asymptotic results (Theorems 3.8.1 and 3.8.2) were established in [17]. The localization result for slow diffusion equation (Theorem 3.6.1) was proved in [16] and the proof presented here closely follows this papers.

In the work [61] the finite extinction is proved with probability one for the self-organized criticality equation (3.63) with the main assumption that $\inf \tilde{\mu} > 0$. The proof relies on sharp estimate on solutions to random differential equation (3.86).

Other extinction results are established in [87]. By the rescaling method presented in Sect. 3.6 were proved in [60] the existence of random attractors for stochastic porous media equations with linearly multiplicative noise.

The finite speed propagation of stochastic porous media equations is also studied in the work [59]. On these lines see also [65].

In [8, 9] an optimal control approach to stochastic porous media equations based on the Brezis–Ekeland variational principle, was developed.

In [45] it is studied the convergence of solutions to stochastic porous media equations when the nonlinearity is convergent in the sense of graphs.

The Sect. 3.6 is based on the work [17]. Previously the rescaling procedure for stochastic porous media equations was used in [24].

Chapter 4
Variational Approach to Stochastic Porous Media Equations

We shall briefly present here a different approach to stochastic porous media equations which in analogy to the variational formulation of parabolic boundary value problems will be called *variational approach*. It is based on a general existence result for infinite dimensional stochastic equations of the form

$$dX(t) + A(t)X(t)dt = \sigma(t, X(t))\, dW(t), \quad X(0) = x,$$

where $A(t) : V \to V^*$ is a family of nonlinear monotone demicontinuous operators and (V, V^*) is a pair of reflexive Banach spaces with a pivot Hilbert space H.

4.1 The General Existence Theory

Let V be a reflexive real Banach space with dual V^* and let H be a separable real Hilbert space such that $V \subset H$ with dense and continuous injection. Then we have

$$V \subset H \subset V^*$$

algebraically and topologically. The duality pairing $_{V^*}(\cdot, \cdot)_V$ coincides with the scalar product $(\cdot, \cdot)$ of H on $H \times H$. Without danger of confusion we shall simply write $(\cdot, \cdot)$ instead of $_{V^*}(\cdot, \cdot)_V$. (Such system (V, H, V') is called a *Gelfand triple*.)

Denote by $\| \cdot \|$ the norm of V and by $| \cdot |$ the norm of H. The norm of V^* which is the dual norm of V, is denoted by $\| \cdot \|_{V^*}$.

Consider the stochastic differential equation

$$\begin{cases} dX(t) + A(t)X(t)dt = \sigma(t, X(t))\, dW(t), & t \in [0, T] \\ \\ X(0) = x \in H, \end{cases} \tag{4.1}$$

© Springer International Publishing Switzerland 2016
V. Barbu et al., *Stochastic Porous Media Equations*, Lecture Notes
in Mathematics 2163, DOI 10.1007/978-3-319-41069-2_4

where $A : [0, T] \times V \times \Omega \rightarrow V^*$ and $\sigma : [0, T] \times V \times \Omega \rightarrow \mathcal{L}_2(Z, H)$ are progressively measurable. Here W is a cylindrical Wiener process in another separable Hilbert space Z defined through a sequence W_j, $j \in \mathbb{N}$, of independent real-valued Brownian motions on a probability space $(\Omega, \mathcal{F}, \mathbb{P})$ with natural filtration $\{\mathcal{F}_t\}_{t \geq 0}$ (see Sect. 1.2.2), and $\mathcal{L}_2(Z, H)$ is the space of Hilbert–Schmidt operators from Z to H.

We further assume that

(i) There is $\lambda > 0$ such that for every $t \in [0, T]$ and $\omega \in \Omega$ the operator $u \rightarrow \lambda u + A(t, \omega)u$ is monotone and demicontinuous from V to V^*. Moreover, $A : [0, T] \times V \times \Omega$ is progressively measurable, that is it is $\mathcal{B}(0, T) \otimes \mathcal{B}(V \otimes \mathcal{F}_t) / \mathcal{B}(V^*)$-measurable.

(ii) There are $1 < p < \infty$ and α_i, $\gamma_i \in \mathbb{R}$, $i = 1, 2, 3$, $\alpha_1 > 0$ such that

$$(A(t, \omega)u, u) \geq \alpha_1 \|u\|_V^p + \alpha_2 |u|_H^2 + \alpha_3, \quad \forall \, u \in V. \tag{4.2}$$

$$\|A(t, \omega)u\|_{V^*} \leq \gamma_1 \|u\|_V^{p-1} + \gamma_2, \quad \forall \, u \in V. \tag{4.3}$$

(iii) σ is Lipschitzian from H to $\mathcal{L}_2(Z, H)$.

In applications to partial differential equations H and V are Sobolev or L^p spaces on domain $\mathcal{O} \subset \mathbb{R}^n$ and $A(t)$, $t \in [0, T]$, is usually an elliptic differential operator on $\mathcal{O}$.

Definition 4.1.1 A continuous $\{\mathcal{F}_t\}_{t \geq 0}$-adapted process $X : [0, T] \rightarrow H$ is called *solution* to (4.1) if X is H-valued pathwise continuous,

$$X \in L^\infty(0, T; L^2(\Omega; H)) \cap L^p((0, T) \times \Omega; V)$$

and $\mathbb{P}$-a.s.

$$X(t) = x - \int_0^t A(s, X(s))ds + \int_0^t \sigma(s, X(s))dW(s), \quad \forall \, t \in [0, T]. \tag{4.4}$$

We have the following existence result [72, 80].

Theorem 4.1.2 *Under hypotheses (i)–(iii) for each $x \in H$ (or more generally for $x \in L^2(\Omega, \mathcal{F}_0, \mathbb{P}, H)$) there is a unique solution $X \in L^2(\Omega; C([0, T]; H))$ to problem* (4.1).

The proof of this important existence theorem can be found in the above cited works and also in [76, 82]. In a few words the idea of the proof is to approximate (4.1) via the Galerkin approximation scheme by a sequence of finite dimensional stochastic differential equations and pass to the limit via monotonicity arguments. For the special case $H = Z = H^{-1}$ and

$$\sigma(X)dW = \sum_{j=1}^\infty \mu_j X f_j \, dW_j,$$

with f_j, $j \in \mathbb{N}$, e.g. as in Sect. 2.1, and $\mu_j \in \mathbb{R}$ such that the series in (4.6) converges absolutely in L^∞, that is, for linear multiplicative noise, a different approach developed in [19] is based on the rescaling transformation $X = e^W y$. In this way Eq. (4.1) is reduced to the random differential equation

$$\begin{cases} \dfrac{dy}{dt} + e^{-W}A(t)(e^W y) + \mu y = 0, & t \in [0, T] \\[2mm] y(0) = x, \end{cases} \tag{4.5}$$

where

$$\mu = \frac{1}{2} \sum_{j=1}^{\infty} \mu_j^2 f_j. \tag{4.6}$$

Equation (4.5) is treated as an operatorial equation of the form $\Lambda y = 0$ where $\Lambda : \mathscr{V} \to \mathscr{V}^*$ is the maximal monotone operator

$$\Lambda y = \frac{dy}{dt} + e^{-W}A(t)(e^W y) + \mu y$$

and $\mathscr{V}$ is the space of all $\{\mathscr{F}_t\}_{t \geq 0}$-adapted processes $X : [0, T] \to V$ such that

$$\mathbb{E} \int_0^T \|e^{W(t)}u(t)\|_V^p \, dt < \infty. \tag{4.7}$$

If we denote by $\mathscr{H}$ the space of all $\{\mathscr{F}_t\}_{t \geq 0}$-adapted processes $u : [0, T] \to H$ such that

$$\mathbb{E} \int_0^T |e^{W(t)}u(t)|^2 \, dt < \infty,$$

with the corresponding norm, then we have

$$\mathscr{V} \subset \mathscr{H} \subset \mathscr{V}^*.$$

(see e.g. [36, p. 278].) This leads to a sharper existence result for (4.1) which implies in particular that the function $t \to e^{-W(t)}X(t)$ is absolutely continuous on $[0, T]$, see [19, Theorem 3.1]. (More will be said about this approach in Sect. 4.4.)

4.2   An Application to Stochastic Porous Media Equations

Consider the stochastic equation

$$\begin{cases} dX(t) - \Delta\psi(t,\xi,X(t))dt = \sigma(t,X(t))dW(t) & \text{in } (0,T)\times\mathscr{O} \\[2mm] X(0,\xi) = x(\xi) & \text{in } \mathscr{O} \\[2mm] \psi(t,\xi,X(t,\xi)) = 0 & \text{on } (0,T)\times\partial\mathscr{O}, \end{cases} \tag{4.8}$$

where $\mathscr{O}$ is a bounded domain in $\mathbb{R}^d$ $d \geq 1$, $\psi : [0,T]\times\overline{\mathscr{O}}\times\mathbb{R}\to\mathbb{R}$ is continuous, monotonically increasing in the third variable, and there exist $a \in (0,\infty)$ and $c \in [0,\infty)$ such that

$$r\psi(t,\xi,r) \geq ar^p - c, \quad \forall\, r \in \mathbb{R}, (t,\xi) \in [0,T]\times\overline{\mathscr{O}},$$

$$\tag{4.9}$$

$$|\psi(t,\xi,r)| \leq c(1 + r^{p-1}), \quad \forall\, r \in \mathbb{R}, (t,\xi) \in [0,T]\times\overline{\mathscr{O}},$$

where $p \in [2d/(d+2),\infty)$ if $d \geq 3$ and $p \in (1,\infty)$ for $d = 1,2$.

Let $\mathscr{H}$ denote the space of all $\{\mathscr{F}_t\}_{t\geq 0}$-adapted processes $u : [0,T]\to H$ such that

$$\mathbb{E}\int_0^T |e^{W(t)}u(t)|^2\,dt < \infty,$$

with the corresponding norm. Then we have

$$\mathscr{V} \subset \mathscr{H} \subset \mathscr{V}^*.$$

By the Sobolev–Gagliardo–Nirenberg theorem, we have $L^p \subset H^{-1}$. To write (4.8) in the form (4.1) we change the pivot space H. Namely, we take $V = L^p$, $H = Z = H^{-1}$, and V^* the dual of V with pivot space H^{-1}. Then $V \subset H \subset V^*$ and

$$V^* = \{\theta \in \mathscr{D}(\mathscr{O}) : \ \theta = -\Delta v, \ v \in L^{p'}\}, \quad p^{-1} + (p')^{-1} = 1,$$

where Δ is taken in the sense of distributions. The duality $_{V^*}(\cdot,\cdot)_V$ is defined as

$$_{V^*}(\theta,u)_V = \int_{\mathscr{O}} \widetilde{\theta}\,u\,d\xi, \quad \widetilde{\theta} = (-\Delta)^{-1}\theta.$$

Here Δ is the Laplace operator with Dirichlet boundary conditions.

The operator $A(t) : V \to V^*$ is defined as

$$_{V^*}(A(t)y, v)_V = \int_{\mathcal{O}} \psi(t, \xi, y(\xi))\, v(\xi)\, d\xi, \quad \forall\, y, v \in V,\ t \in [0, T].$$

By (4.9) we infer that $A(t)$ satisfies (4.2), (4.3), that is

$$_{V^*}(A(t)y, y)_V \geq \alpha_1 \|y\|_V^p + \alpha_2 \quad \forall\, y \in V,$$

$$\|A(t, y)\|_{V^*} \leq \gamma_1 \|y\|_V^{p-1} + \gamma_2, \quad \forall\, y \in V.$$

It is also readily seen that $A(t) : V \to V^*$ is demicontinuous.

Now applying Theorem 4.1.2 to the present situation we get

Theorem 4.2.1 *For each $x \in H^{-1}$ there is a unique distributional solution*

$$X \in L^2(\Omega; C([0, T]; H^{-1}) \cap L^2((0, T) \times \Omega \times \mathcal{O})$$

to Eq. (4.8).

Remark 4.2.2 The analysis of corresponding rescaling Eq. (4.5) reveals that the solution to (4.8) is of the form $X = e^W y$ where $y = y(t, \omega)$ is the solution to (4.5) which is, for each $\omega \in \Omega$, V^*-absolutely continuous on $[0, T]$ and

$$\mathbb{E} \int_0^T \left\| e^{W(t)} \frac{d}{dt} y(t) \right\|_{V^*}^{p'} dt < \infty.$$

(See [19, Corollary 6.8]).

Remark 4.2.3 It should be noted that assumption (4.9) covers the *slow diffusions* case only and excludes some *fast diffusions* equations of the form (4.8). In order to treat the latter case via abstract Theorem 4.1.2 one should replace the space $V = L^p$ by an Orlicz space which will be done in the next section (see [84]).

4.3 Stochastic Porous Media Equations in Orlicz Spaces

Let us briefly review definition and basic properties of Orlicz spaces (see [1]). Let $N : \mathbb{R} \to \mathbb{R}$ be a nonnegative, continuous, even, convex function on $\mathbb{R}^+$ such that $N(s) = 0$ iff $s = 0$ and

$$\lim_{s \to 0} \frac{N(s)}{s} = 0, \quad \lim_{s \to \infty} \frac{N(s)}{s} = +\infty,$$

$$\frac{N(s)}{s} > \frac{N(t)}{t} \quad \text{if } s > t > 0.$$

The Orlicz space L_N is by definition the space of all measurable functions $u : \mathcal{O} \to \mathbb{R}$ such that

$$\int_{\mathcal{O}} N(|u(x)|) \, ds < \infty.$$

In general L_N is not linear but this happens if the Young function N is Λ_2-regular, that is if there is $C > 0$ such that

$$N(2r) \le CN(r), \quad \forall \, r \ge 0. \tag{4.10}$$

It turns out that L_N in this case is a Banach space with the norm (Luxemburg norm)

$$\inf \left\{ \lambda > 0 : \int_{\mathcal{O}} N \left(\frac{|u(x)|}{\lambda} \right) dx \le 1 \right\} =: \|u\|_{L^N}. \tag{4.11}$$

Let N^* denote the conjugate of N, that is (see 1.2.4)

$$N^*(r) = \sup_s \{rs - N(s)\}.$$

It is a Young function too.

If N and its conjugate N^* are Λ_2-regular then L_N and L_{N^*} are Banach spaces with the duality

$$_{L_N^*}(f, g)_{L_N} = \int_{\mathcal{O}} f(x)g(x)dx$$

and so L_N is a reflexive Banach space (see [1, p. 237]). Moreover, one has

$$\|f\|_{L_N^*} \, \|g\|_{L_N} \le \left| _{L_N^*} (f, g)_{L_N} \right| \le 2 \|f\|_{L_N^*} \, \|g\|_{L_N}.$$

We recall also that (see [1, p. 234]) $L_{N_1} \subset L_{N_2}$ if N_1 dominates N_2 at ∞, that is there is a constant $k > 0$ such that

$$N_2(r) \le N_1(kr), \quad \forall \, r \ge 0.$$

N_1 and N_2 are equivalent at ∞ if each dominates the other, that is

$$0 < \lim_{r \to \infty} \frac{N_1(r)}{N_2(r)} < \infty.$$

In particular, we have

$$L^p \subset L_N, \quad 1 \le p < \infty$$

if

$$\lim_{r \to \infty} \frac{r^p}{N(r)} > 0.$$

This implies via the Sobolev embedding theorem that $H_0^1 \subset L^{p^*} \subset L_{N^*}$ and so $L_N \subset H^{-1}$ if the function $r \to r^{p^*}$, which is of course a Young function, dominates N^*, that is

$$\lim_{r \to \infty} \frac{r^{p^*}}{N^*(r)} > 0, \tag{4.12}$$

where p^* is the Sobolev index,

$$p^* = \frac{2d}{d-2} \quad \text{if } d > 2, \quad p^* \in [2, \infty) \quad \text{if } d = 1, 2. \tag{4.13}$$

We note also that by the Λ_2-regularity of N and N^* we have

$$N(\lambda r) \le K(\lambda) N(r), \quad \forall \lambda > 0, \ r \ge 0, \tag{4.14}$$

$$N^*(\lambda r) \le K^*(\lambda) N^*(r), \quad \forall \lambda > 0, \ r \ge 0, \tag{4.15}$$

where K and K^* are continuous, positive and monotonically increasing functions. We shall also assume that there is $1 < q < \infty$ such that

$$(K^*)^{-1}(K(r)) \le C_1 r^q + C_2, \quad \forall r \ge 0. \tag{4.16}$$

$$|K(r)| \ge \gamma_1 r^{q+1} + \gamma_2, \quad \forall r \ge 0. \tag{4.17}$$

Let us give below a few examples of N-functions along with their conjugates.

Example 4.3.1

(1) $N(r) = \frac{1}{p} r^p$, $N^*(r) = \frac{1}{p'} r^{p'}$, $p^{-1} + (p')^{-1} \doteq 1$, $1 < p < \infty$.
(2) $N(r) = e^r - r - 1$, $N^*(r) = (1 + r) \log(1 + r) - r$.
(3) $N(r) = r^{\alpha_1} \log(1 + r)^{\alpha_2}$, $\alpha_1 > 1$, $\alpha_2 \ge 1$.

It is easily seen in the later case that $N^*(r) = o(r^{1+\delta})$, $\delta > 0$, for $r \to \infty$ and that condition (4.12) holds. Also in this case conditions (4.16) and (4.17) holds for some $q = \alpha_1 - \epsilon$, ϵ arbitrary small.

We are going to represent Eq. (4.8) in the abstract setting (4.1) with V, H and $A(t)$ suitably chosen. To this purpose we take

$$H = H^{-1}, \quad V = H^{-1} \cap L_N \tag{4.18}$$

and $A(t) : V \to V^*$ defined by

$$_V\langle A(t)u, v\rangle_{V^*} = \int_{\mathcal{O}} \psi(t, \xi, u(\xi)) \, v(\xi) \, d\xi, \quad \forall \, u, v \in L_N. \tag{4.19}$$

Here V^* is the dual space of V with the pivot space H^{-1} (that is $_{V^*}\langle v^*, v\rangle_V = \langle v^*, v\rangle_{-1}$ for $v^* \in H^{-1}$, $v \in V$) and we have

$$V \subset H \subset V^* \tag{4.20}$$

in the algebraic and topological sense. Of course the norm $\|\cdot\|_V$ on V is taken as

$$\|u\|_V = \|u\|_{-1} + \|u\|_{L_N}. \tag{4.21}$$

If (4.12), (4.13) hold, and for the sake of simplicity we shall assume this in the following, then as seen above $L_N \subset H^{-1}$ and so $V = L_N$ (see [84] for a complete exposition of the general case). Moreover, the space V^* is given by

$$V^* = \{\theta \in \mathscr{D}'(\mathcal{O}) : \theta = -\Delta u, \, u \in L_{N^*}\}.$$

We assume that $\psi : [0, T] \times \mathcal{O} \times \mathbb{R} \to \mathbb{R}$ satisfies the following conditions

$$\psi(t, \xi, 0) = 0, \, (\psi(t, \xi, r) - \psi(t, \xi, s))(r - s) \geq 0, \, \forall \, r, s \in \mathbb{R}, \, \xi \in \mathcal{O}, \tag{4.22}$$

$$r(\psi(t, \xi, r) \geq N(r), \, \forall \, t \in [0, T], \, r \in \mathbb{R}, \, \xi \in \mathcal{O}, \tag{4.23}$$

$$C(N(r) + 1) \geq r\psi(t, \xi, r) \geq N(r), \, \forall \, t \in [0, T], \, r \in \mathbb{R}, \, \xi \in \mathcal{O}, \tag{4.24}$$

where $C > 0$.

We note first that by (4.22)–(4.24) it follows that the operator $A(t)$ is monotone from V to V^* and demicontinuous for each $t \in [0, T]$. Indeed, by (4.24) we have for all $r \geq 0$

$$N^*(C^{-1}\psi(t, \xi, r)) = \sup_s\{C^{-1}s\psi(t, \xi, r) - N(s)\}$$

$$\leq C^{-1} \sup_{s \geq 0}\{s\psi(t, \xi, r) - \psi(t, \xi, s)\} + C$$

$$\leq C^{-1} \sup_{r \geq s \geq 0} \{s(\psi(t, \xi, r) - \psi(t, \xi, s))\} + C \tag{4.25}$$

$$\leq C^{-1}r\psi(t, r) + C \leq N(r) + C.$$

This implies that for all $u \in L_N$, $\psi(u) \in L_N$ and

$$_{V^*}\langle A(t)u, v \rangle_V = \int_{\mathcal{O}} \psi(t, \xi, u(\xi)) \, v(t, \xi) d\xi$$

$$\leq 2 \|\psi(u)\|_{L_N^*} \|v\|_{L_N}$$

$$\leq 2 \|v\|_{L_N} \inf \left\{ \lambda > 0 : \int_{\mathcal{O}} N^* \left(\frac{|\psi(t, \xi, u(\xi))|}{\lambda} \right) d\xi \leq 1 \right\}.$$

$$(4.26)$$

On the other hand, by (4.18)–(4.19) and (4.25) we have

$$\inf \left\{ \lambda > 0 : \int_{\mathcal{O}} N^* \left(\frac{|\psi(t, \xi, u(\xi))|}{\lambda} \right) d\xi \right\}$$

$$\leq \inf \left\{ \lambda > 0 : K^* \left(\frac{C}{\lambda} \right) \int_{\mathcal{O}} N^* (C^{-1} \psi(t, \xi, u(\xi))|) \, d\xi \leq 1 \right\}$$

$$\leq \inf \left\{ \lambda > 0 : K^* \left(\frac{C}{\lambda} \right) \int_{\mathcal{O}} N(|u(\xi)) \, d\xi + C|\mathcal{O}|) \leq 1 \right\}$$

$$\leq \inf \left\{ \lambda > 0 : K^* \left(\frac{C}{\lambda} \right) K(\|u\|_{L_N} \int_{\mathcal{O}} N \left(\frac{|u(\xi)|}{\|u\|_{L_N}} \right) d\xi + C|\mathcal{O}|) \leq 1 \right\}$$

Since by (4.11)

$$\int_{\mathcal{O}} N \left(\frac{|u(\xi)|}{\|u\|_{L_N}} \right) d\xi < 1$$

we get

$$\inf \left\{ \lambda > 0 : \int_{\mathcal{O}} N^* \left(\frac{|\psi(t, \xi, u(\xi))|}{\lambda} \right) d\xi \right\} \leq \lambda^*(u), \tag{4.27}$$

where $\lambda^*(u)$ is given by

$$K^* \left(\frac{C}{\lambda^*(u)} \right) K(\|u\|_{L_N} + C|\mathcal{O}|) = 1,$$

that is

$$\lambda^*(u) = C \left((K^*)^{-1} \left(\frac{1}{K(\|u\|_{L_N}) + C|\mathcal{O}|)} \right) \right)^{-1}. \tag{4.28}$$

Then by (4.26)–(4.28) we have

$$\|A(t)u\|_{V^*} \le C\left((K^*)^{-1}\left(\frac{1}{K(\|u\|_{L_N})+C|\mathcal{O}|}\right)\right)^{-1}$$

$$\le C_3(K^*)^{-1}K(\|u\|_{L_N}) + C_4, \quad \forall\, u \in V,$$

for $C_3, C_4 > 0$ independent of u. Then by assumption (4.16) we have

$$\|A(t)u\|_{V^*} \le C_5\|u\|_V^q + C_6, \quad \forall\, u \in V, \tag{4.29}$$

where $C_5, C_6 > 0$ and $q \in (1,\infty)$.

It is also easily seen by (4.26) that $A(t)$ is for all $t \in [0,T]$ and $\omega \in \Omega$ demicontinuous from V to V^*. By (4.23) and (4.26) we have also

$$_{V^*}\langle A(t)u, u\rangle_V = \int_{\mathcal{O}} \psi(t,\xi,u(\xi))\,u(\xi)\,d\xi$$

$$\ge \int_{\mathcal{O}} N(|u(\xi)|)\,d\xi = \int_{\mathcal{O}} N\left(\frac{u(\xi)}{\|u\|_{L_N}}\right)d\xi \tag{4.30}$$

$$\ge K(\|u\|_{L_N}), \quad \forall\, u \in V.$$

Then, if we assume that (4.10), (4.12), (4.13) hold, by assumption (4.17)

$$_{V^*}\langle A(t)u, u\rangle_V \ge \gamma_1\|u\|_{L_N}^{q+1} + \gamma_2 \ge \gamma_1\|u\|_V^{q+1} + \gamma_2, \quad \forall\, u \in V.$$

Then we may apply Theorem 4.1.2 for $p = q+1$ and $A(t)$, V, H as above and obtain.

Theorem 4.3.2 *Assume that Hypotheses* (4.10), (4.12), (4.13), (4.16), (4.17), (4.22)–(4.24) *hold. Then for each* $x \in H^{-1}$ *there is a unique distributional solution*

$$X \in L^2(\Omega; C([0,T]; H^{-1})) \cap L^2((0,T) \times \Omega; L_N))$$

to Eq. (4.8).

In particular, Theorem 4.3.2 can be applied for fast diffusion equations (4.8) where

$$\psi(r) = \operatorname{sign} |r|^{\alpha_1 - 1} \log(1 + |r|)^{\alpha_2}, \quad \forall\, r \in \mathbb{R}, \tag{4.31}$$

where $\alpha_1 \in (1,\infty)$, $\alpha_2 \in (1,\infty)$. In this case assumptions (4.23), (4.24) hold for the Young function N defined by Example 4.3.1(3). In fact in this case Theorem 4.3.2 covers for $d \ge 3$ some cases which are not included in Theorem 4.1.2.

4.4 Comments and Bibliographical Remarks

The use of Orlicz spaces is quite familiar in the existence theory of PDEs with nonlinear terms having non polynomial growth. On the other hand, it is apparent that assumptions (4.10) and (4.16), (4.17) are quite restrictive for the class of Orlicz spaces L_N where the problem was treated. In fact, roughly speaking, this means that the Young functions N and N^* are comparable at infinity with polynomial functions and so, are not "far away" from spaces L^p. The main reason for this restriction is the use of Theorem 4.1.2 which involves polynomial growth for $A(t)u$. In [84] one uses a direct approach (via Galerkin scheme) for Eq. (4.1) avoiding assumptions (4.12), (4.13), (4.16) and (4.17), with $A(t) : V \to V^*$ defined by (4.30) under assumption

$$\langle A(t)u, u \rangle_V \geq -C_1 \|u\|_{-1}^2 + R(u), \quad \forall u \in V,$$

$$\|A(t)u\|_{V^*} \leq C_2(R(u) + 1),$$

where $R : V \to [0, \infty)$ is such that $\mathbb{E} \int_0^T R(z(t))dt$ is "comparable" (in a certain sense) with the norm $\|z\|_K$ of a reflexive Banach space K where

$$L^p((0, T) \times \Omega; V) \subset K \subset L^1((0, T) \times \Omega; V), \quad 1 < p < \infty.$$

(See the above cited papers [84, 86] for a precise definition of function R.)

However, even in this case the necessary condition on Eq. (4.8) precludes the diffusion functions with exponential growth as well as the superfast diffusions, which will be treated in Chap. 5 by a different method.

In order to exploit completely the generality offered by the Orlicz spaces one should extend the basic Theorem 4.3.2 allowing nonlinear monotone demicontinuous operators $A(t) : V \to V^*$ which instead of (i), (ii) satisfy the following conditions

$$(A(t, \omega)u, u) \geq N_1(\|u\|_V), \quad \forall u \in V, \ t \in [0, T], \tag{4.32}$$

$$\|A(t, \omega)u\|_{V^*} \leq N_2(\|u\|_V), \quad \forall u \in V, \ t \in [0, T], \tag{4.33}$$

where N_1, N_2 are two Young functions satisfying

$$N_1(r) \leq N_2(r), \quad \forall r \in \mathbb{R}$$

In the case of linear multiplicative noise an existence result for a solution X to (4.1) satisfying $X \in L^\infty((0, T) \times \Omega; H)$ and

$$\mathbb{E} \int_0^T N_2(\|X(t)\|_V)dt < \infty$$

might be obtained along the lines developed in [19] by representing the rescaled random equation (4.5) as an operatorial equation $\mathscr{B}y + \mathscr{A}y = 0$ where the operators $\mathscr{A} : \mathscr{V} \to \mathscr{V}^*, \mathscr{B} : \mathscr{V} \to \mathscr{V}^*$ are given by

$$(\mathscr{A}y)(t) = e^{-W(t)}A(t)e^{W(t)}y(t), \quad \mathscr{B}y(t) = \frac{dy}{dt} + \mu y$$

and $\mathscr{V}$ is the space of $\{\mathscr{F}_t\}_{t \geq 0}$-adapted processes $u : [0, T] \to V$ such that

$$\mathbb{E} \int_0^T N_1(\|e^{W(t)} u(t)\|_V)dt < \infty.$$

Then, under assumptions (4.32)–(4.33), $\mathscr{A}$ is maximal monotone coercive from $\mathscr{V}$ to $\mathscr{V}^*$ and everywhere defined, while the operator $\mathscr{B}$ with domain

$$D(\mathscr{B}) = \left\{ y \in V : \frac{dy}{dt} \in \mathscr{V}^*, \, y(0) = x \right\},$$

is maximal monotone from $\mathscr{V}$ to $\mathscr{V}^*$ and hence $R(\mathscr{A} + \mathscr{B}) = \mathscr{V}^*$. The proof is as in [19] but we omit the details.

We also note that Theorem 4.3.2 extends to stochastic porous media equation (4.8) on unbounded domain $\mathcal{O}$ and in particular for $\mathcal{O} = \mathbb{R}^N$ and we refer to [84] for the treatment in this case. (See also [86]). (The latter case will be also treated in Chap. 6.)

Chapter 5
L^1-Based Approach to Existence Theory for Stochastic Porous Media Equations

The existence theory developed in the previous chapter was based on energy estimates in the space H^{-1} obtained via Itô's formula in approximating equations. This energetic approach leads to sharp existence results, but requires polynomial growth assumptions or strong coercivity for the nonlinear function β. The case of general maximal monotone functions β of arbitrary growth and in particular with exponential growth was beyond the limit of the previous theory. Here we develop a different approach based on sharp L^1-estimates for the corresponding approximating equations which allows to treat these general situations.

5.1 Introduction and Setting of the Problem

We are here concerned with the equation

$$\begin{cases} dX(t) = \Delta\beta(X(t))dt + \sigma(X(t))dW(t), \\ x(0) = x \in H^{-1}, \end{cases} \tag{5.1}$$

where β is a maximal monotone graph in $\mathbb{R}\times\mathbb{R}$. In Chap. 3 this problem was studied under Hypothesis 4(i). Here the existence theory is extended to general multivalued maximal monotone graphs β such that $\beta(\mathbb{R}) = \mathbb{R}$.

More precisely, we shall assume that

Hypothesis 7

(i) $\beta : \mathbb{R} \to 2^{\mathbb{R}}$ is a maximal monotone graph such that $0 \in \beta(0)$, $D(\beta) = \beta(\mathbb{R}) = \mathbb{R}$, and

$$\limsup_{|s|\to+\infty} \frac{j(-s)}{j(s)} < +\infty, \tag{5.2}$$

© Springer International Publishing Switzerland 2016
V. Barbu et al., *Stochastic Porous Media Equations*, Lecture Notes
in Mathematics 2163, DOI 10.1007/978-3-319-41069-2_5

where $j : \mathbb{R} \to \mathbb{R}$ is the potential of β, i.e. $\partial j = \beta$.
(ii) σA^γ is Lipschitzian from H^{-1} to $\mathscr{L}_2(H^{-1})$ where $\gamma > d/2$.
(iii) $W(t)$ is a cylindrical Wiener process on H^{-1} of the form (2.10).

We shall denote by $j^* : \mathbb{R} \to \mathbb{R}$ the conjugate (the Legendre transform) of j,

$$j^*(p) = \sup\{py - j(y) : y \in \mathbb{R}\}.$$

We recall that $\partial j^* = (\partial j)^{-1}$ (see (1.38)–(1.39)),

$$j(y) + j^*(p) = py \quad \text{if and only if } p \in \partial j(y) \tag{5.3}$$

and

$$j(u) + j^*(p) \geq pu \quad \text{for all } p, u \in \mathbb{R}. \tag{5.4}$$

Remark 5.1.1 We note that as $D(\beta) = \mathbb{R}$, the convex function j is continuous. Moreover, since $0 \in \beta(0)$, we have $j(0) = \inf j$. Hence subtracting $j(0)$ we can take j such that $j(0) = 0$ and $j \geq 0$ and therefore we may assume that $j^* \geq j^*(0) = 0$. We recall (see e.g. [6, page 80], [35]) that the condition $R(\beta) = \mathbb{R}$ is equivalent to

$$j^*(y) < \infty, \ \forall \, y \in \mathbb{R}, \quad \lim_{|y| \to \infty} \frac{j(y)}{|y|} = +\infty. \tag{5.5}$$

while the condition $D(\beta) = \mathbb{R}$ is equivalent to

$$j(y) < \infty, \ \forall \, y \in \mathbb{R} \iff \lim_{|y| \to \infty} \frac{j^*(y)}{|y|} = +\infty.$$

Hypothesis 7(i) automatically holds if β is a monotonically increasing, continuous function on $\mathbb{R}$ satisfying the conditions

$$\limsup_{|s| \to +\infty} \frac{\int_0^{-s} \beta(t)dt}{\int_0^{s} \beta(t)dt} < +\infty. \tag{5.6}$$

and

$$\lim_{s \to +\infty} \beta(s) = +\infty, \quad \lim_{s \to -\infty} \beta(s) = -\infty.$$

In particular, it is satisfied by functions β of the form

$$\beta(s) = a \log(s^{2k} + 1) \, \text{sgn} \, s, \quad s \in \mathbb{R}. \tag{5.7}$$

for $a, k > 0$ or more generally by those satisfying assumption $(A1)$ in [84].

Another case not covered by Chap. 3 but which is within Hypothesis 7 is that of β with exponential growth to $\pm\infty$. For instance

$$\beta(s) := a|s|^{2p-1} s \, \exp\{b|s|^{2m}\},$$

where $p, m \geq 1$, $a > 0$.

We set $j_\lambda(u) = \int_0^u \beta_\lambda(r)dr$ and recall that it is equal to the Moreau approximation of j, i.e., (see (1.36), (1.37))

$$j_\lambda(u) = \min\left\{ j(v) + \frac{1}{2\lambda}\,|u - v|^2 : v \in \mathbb{R} \right\}. \tag{5.8}$$

$$j_\lambda(u) = j((1 + \lambda\beta)^{-1}u) + \frac{1}{2\lambda}\,|u - (1 + \lambda\beta)^{-1}u|^2. \tag{5.9}$$

Definition 5.1.2 A process $X \in C_W([0, T]; H^{-1}) \cap L^1((0, T) \times \mathcal{O} \times \Omega)$, such that $X \in C^w([0, T], H^{-1})$, $\mathbb{P}$-a.s., is said to be a distributional solution to Eq. (5.1) if there exists a process $\eta \in L^1((0, T) \times \mathcal{O} \times \Omega)$ such that

$$\eta\,(t, \xi) \in \beta(X(t, \xi)), \quad \text{a.e. } (t, \xi) \in Q_T, \ \mathbb{P}\text{-a.s.} \tag{5.10}$$

$$\int_0^\bullet \eta\,(s)ds \in C^w([0, T]; H_0^1), \tag{5.11}$$

$$X(t) - \Delta \int_0^t \eta(s)ds = x + \int_0^t B(X(s))dW(s), \quad \forall\, t \in [0, T], \ \mathbb{P}\text{-a.s.} \tag{5.12}$$

$$j(X), j^*\,(\eta) \in L^1((0, T) \times \mathcal{O} \times \Omega). \tag{5.13}$$

(Here $\int_0^t \eta(s)ds$ is initially defined as on L^1-valued Bochner integral). Of course, if β is single valued then (5.10)–(5.12) reduce to

$$\int_0^\bullet \beta(X(s))ds \in C^w([0, T]; H_0^1), \tag{5.14}$$

and

$$X(t) - \Delta \int_0^t \beta(X(s))ds = x + \int_0^t B(X(s))dW(s), \quad \forall\, t \in [0, T], \ \mathbb{P}\text{-a.s.}. \tag{5.15}$$

Here $C^w([0, T]; H^{-1})$ is the space of weakly continuous functions $y : [0, T] \to H^{-1}$ and the space $C^w([0, T]; L^2)$ is similar defined. The Laplace operator Δ is considered in the sense of distributions on $\mathcal{O}$.

Definition 5.1.2 is related to Definition 3.1.2 of strong or distributional solutions for Eq. (1.1). We note that X, as in Definition 5.1.2, is automatically $(\mathscr{F}_t)_{t\geq 0}$-adapted.

Theorem 5.1.3 is the main result of this chapter.

Theorem 5.1.3 *Under Hypothesis 7, for each $x \in H^{-1}$ there is a unique distributional solution $X = X(t, x)$ to Eq. (5.1). Moreover, the following estimate holds*

$$\mathbb{E}\|X(t,x) - X(t,y)\|^2_{-1} \leq C\|x - y\|^2_{-1}, \quad \text{for all } t \geq 0, \tag{5.16}$$

where C is independent of $x, y \in H$.

Before proving Theorem 5.1.3 by a fixed point argument, we shall establish the existence of solutions for the equation

$$\begin{cases} dY(t) - \Delta\beta(Y(t))dt = G(t)dW(t) & \text{in } Q_T, \\ \beta(Y(t)) = 0 & \text{on } \Sigma_T, \\ Y(0) = x & \text{in } \mathscr{O}, \end{cases} \tag{5.17}$$

where $G : [0, T] \to \mathscr{L}_2(L^2, D(A^\gamma))$ is an $(\mathscr{F}_t)_{t\geq 0}$-adapted process such that

$$\mathbb{E}\int_0^T \|G(t)\|^2_{\mathscr{L}_2(L^2, D(A^\gamma))}dt < +\infty \tag{5.18}$$

and $\gamma > d/2$. Here GW is given by

$$GW = \sum_{k=1}^\infty Gf_k W_k.$$

We notice that by (5.18) it follows that $G \in L^2(0, T; H_0^1)$ $\mathbb{P}$-a.s.

A solution of (5.17) is defined to be an $(\mathscr{F}_t)_{t\geq 0}$-adapted H^{-1}-valued process Y satisfying along with $\eta \in L^1((0, T) \times \mathscr{O} \times \Omega)$ Hypothesis 7 where $\sigma(X)$ is replaced by G.

Theorem 5.1.4 *Under Hypothesis 7 for each $x \in H^{-1}$ there is a unique distributional solution $Y = Y_G(t, x)$ to Eq. (5.17). Moreover, the following estimate holds*

$$\mathbb{E}\|\dot{Y}_{G_1}(t, x) - Y_{G_2}(t, y)\|^2_{-1} \leq \|x - y\|^2_{-1}$$

$$+ \mathbb{E}\int_0^t \|G_1(s) - G_2(s)\|^2_{\mathscr{L}_2(L^2, H^{-1})}ds, \tag{5.19}$$

for all $t \geq 0$, $x, y \in H^{-1}$ and G_1, G_2 satisfying (5.18).

Remark 5.1.5 It should be noted that Hypothesis 7(ii) excludes the case where the covariance operator σ is of the form $\sigma(x) \equiv x$, i.e. the case of linear multiplicative noise. This case, however, can be treated too for special Wiener processes of the form $W = \sum_{k=1}^{N} \mu_k f_k W_k$ by the methods developed here.

Remark 5.1.6 Hypothesis 7(iii) for example allows monotonically increasing functions β which are continuous from the right on $\mathbb{R}$ and have a finite number of jumps $r_1, r_2, \ldots, r_N$. However in this case one must fill the jumps by replacing the function β by the maximal monotone (multivalued) graph $\tilde{\beta}(r) = \beta(r)$ for $r \neq r_i$ and $\tilde{\beta}(r_i) = [\beta(r_i) - \beta(r_{i-1} - 0)]$. Such a situation might arise in modeling underground water flows (see e.g. [79]). In this case β is the diffusivity function and (5.1) reduces to *Richard's equation*. It must be also said that Theorems 5.1.3 and 5.1.4 have natural extensions to equations of the form

$$dX(t) - \Delta\beta(X(t))dt + \Phi(X(t))dt = \sigma(X(t))dW(t), \qquad (5.20)$$

where Φ is a suitable monotonically increasing and continuous function (see [84]).

5.2 Proof of Theorem 5.1.4

Following the procedure developed in the previous case, we consider the approximating equation

$$\begin{cases} dX_\lambda(t) - \Delta(\beta_\lambda(X_\lambda(t)) + \lambda X_\lambda(t))dt = G(t)dW(t) & \text{in } (0, T) \times \mathcal{O} := Q_T, \\[2mm] \beta_\lambda(X_\lambda(t)) + \lambda X_\lambda(t)) = 0 & \text{on } (0, T) \times \partial\mathcal{O}, \\[2mm] X_\lambda(0) = x & \text{in } \mathcal{O}, \end{cases}$$

$$(5.21)$$

which, for each $\lambda > 0$ and $x \in L^2$, has a unique solution $X_\lambda \in C_W([0, T]; H^{-1}) \cap L^2(\Omega; C([0, T]; H^{-1}))$ such that $X_\lambda, \beta_\lambda(X_\lambda) \in L^2_W(0, T; H^1_0)$.

To get this one might use the results from Chap. 2 but it is more convenient to give a direct argument which works for stochastic differential equations with additive noise.

Indeed, setting $y_\lambda(t) = X_\lambda(t) - W_G(t)$ where $W_G(t) = \int_0^t G(s)dW(s)$, we may rewrite (5.21) as the random differential equation

$$\begin{cases} y_\lambda'(t) - \Delta\tilde{\beta}_\lambda(y_\lambda(t) + W_G(t)) = 0 & \mathbb{P}\text{-a.s. in } Q_T, \\[2mm] \tilde{\beta}_\lambda(y_\lambda(t) + W_G(t)) = 0 & \text{on } (0, T) \times \partial\mathcal{O}, \\[2mm] y_\lambda(0) = x & \text{in } \mathcal{O}, \end{cases}$$

$$(5.22)$$

where $\tilde{\beta}_\lambda(y) = \beta_\lambda(y) + \lambda y$, $\lambda > 0$. Note that $\tilde{\beta}_\lambda(0) = 0$.

For each $\omega \in \Omega$ and $t \in [0, T]$ the operator $\Gamma(t) : H_0^1 \to H^{-1}$, defined by

$$\Gamma(t)y = -\Delta\tilde{\beta}_\lambda(y + W_G(t)), \quad y \in H_0^1,$$

is continuous, monotone and coercive, i.e.,

$$(\Gamma(t)y, y) \geq \lambda\|y + W_G(t)\|_1^2 - (\Gamma(t)y, W_G(t))$$

$$\geq \frac{\lambda}{2}\|y\|_1^2 - C_\lambda\|W_G(t)\|_1^2,$$

where $\|\cdot\|_1$ is as usually the norm of H_0^1. Then by classical existence theory for infinite dimensional Cauchy problems (see e.g. [74, Page 170]) Eq. (5.22) has a unique solution

$$y_\lambda = y_\lambda(t, x) \in C([0, T]; L^2) \cap L^2(0, T; H_0^1)$$

with $y'_\lambda \in L^2(0, T; H^{-1})$, that is $y_\lambda \in W^{1,2}([0, T]; H^{-1})$.

By (5.22) it is readily seen that

$$\|y_\lambda(t, x) - y_\lambda(t, \bar{x})\|_{-1} \leq C\|x - \bar{x}\|_{-1}, \quad \forall\, x, \bar{x} \in H^{-1}$$

and so, $y_\lambda(t, x)$ extends as a solution $y_\lambda \in C([0, T]; H^{-1} \cap L^2(0, T; L^2)$ of (5.22) for all $x \in H^{-1}$.

5.2.1   A-Priori Estimates

We fix $\omega \in \Omega$ and work with the corresponding solution y_λ to (5.22). It suffices to assume that $x \in L^2$. We have by (5.22)

$$\frac{1}{2}\frac{d}{dt}\|y_\lambda(t)\|_{-1}^2 + \langle\tilde{\beta}_\lambda(y_\lambda(t) + W_G(t)), y_\lambda(t) + W_G(t)\rangle_2$$

$$= \langle\tilde{\beta}_\lambda(y_\lambda(t) + W_G(t)), W_G(t)\rangle_2,$$
(5.23)

which is equivalent to

$$\frac{1}{2}\frac{d}{dt}\|y_\lambda(t)\|_{-1}^2 + \langle\beta_\lambda(y_\lambda(t) + W_G(t)), y_\lambda(t) + W_G(t)\rangle_2$$

$$= -\lambda\langle y_\lambda(t), y_\lambda(t) + W_G(t)\rangle_2 + \langle\beta_\lambda(y_\lambda(t) + W_G(t)), W_G(t)\rangle_2.$$
(5.24)

Now set $j_\lambda(u) = \int_0^u \beta_\lambda(r)dr$ and let j_λ^* denote the conjugate of j_λ. Moreover we set

$$z_\lambda = (1 + \lambda\beta)^{-1}(y_\lambda + W_G), \quad \eta_\lambda = \beta_\lambda(y_\lambda + W_G). \tag{5.25}$$

The aim of this section is to prove the following estimate: there exists a (random) constant C_1 such that

$$\frac{1}{2}\|y_\lambda(t)\|_{-1}^2 + \int_0^t \int_{\mathcal{O}} (j(z_\lambda) + j^*(\eta_\lambda))d\xi ds$$

$$+\frac{1}{2\lambda} \int_0^t \int_{\mathcal{O}} (y_\lambda + W_G - z_\lambda)^2 d\xi ds \le C_1(1 + \|x\|_{-1}^2), \quad t \in [0, T]. \tag{5.26}$$

By (5.3) we have

$$j_\lambda^*(\beta_\lambda(y_\lambda(t) + W_G(t))) + j_\lambda(y_\lambda(t) + W_G(t)) = \beta_\lambda(y_\lambda(t) + W_G(t))(y_\lambda(t) + W_G(t)). \tag{5.27}$$

Substituting this identity into (5.24) yields

$$\frac{1}{2}\|y_\lambda(t)\|_{-1}^2 + \int_0^t \int_{\mathcal{O}} (j_\lambda(y_\lambda(s) + W_G(s)) + j_\lambda^*(\beta_\lambda(y_\lambda(s) + W_G(s))))d\xi ds$$

$$= \frac{1}{2}\|x\|_{-1}^2 + \int_0^t \int_{\mathcal{O}} (\beta_\lambda(y_\lambda(s) + W_G(s))W_G(s))d\xi ds$$

$$-\lambda \int_0^t \int_{\mathcal{O}} y_\lambda(s)(y_\lambda(s) + W_G(s))d\xi ds, \tag{5.28}$$

Then, using (5.9) and the fact that $j_\lambda^* \ge j^*$ for all $\lambda > 0$, we see that

$$\frac{1}{2}\|y_\lambda(t)\|_{-1}^2 + \int_0^t \int_{\mathcal{O}} (j(z_\lambda(s)) + j^*(\eta_\lambda(s)))d\xi ds$$

$$+\frac{1}{2\lambda} \int_0^t \int_{\mathcal{O}} (y_\lambda(s) + W_G(s) - z_\lambda(s))^2 d\xi ds$$

$$\le \frac{1}{2}\|x\|_{-1}^2 + \int_0^t \int_{\mathcal{O}} \eta_\lambda(s)W_G(s)d\xi ds - \lambda \int_0^t \int_{\mathcal{O}} y_\lambda(s)(y_\lambda(s) + W_G(s))d\xi ds. \tag{5.29}$$

We now estimate the first integral from the right hand side of (5.29) as follows

$$\left| \int_0^t \int_{\mathscr{O}} \eta_\lambda(s) W_G(s) d\xi ds \right| \le \delta \int_0^t \int_{\mathscr{O}} |\eta_\lambda(s)| d\xi ds, \tag{5.30}$$

where $\delta := \sup_{s \in [0,T]} |W_G(s)|_{L^\infty} < +\infty$. We note that by assumption (5.18) and since $\gamma > d/2$ it follows by Sobolev embedding that $W_G(\cdot)$ has continuous sample paths in $D(A^\gamma) \subset L^\infty$ and so δ is indeed finite.

Substituting (5.30) in (5.29) yields

$$\frac{1}{2} \|y_\lambda(t)\|_{-1}^2 + \int_0^t \int_{\mathscr{O}} (j(z_\lambda(s)) + j^*(\eta_\lambda(s)) d\xi ds$$

$$+ \frac{1}{2\lambda} \int_0^t \int_{\mathscr{O}} (y_\lambda(s) + W_G - z_\lambda(s))^2 d\xi ds$$

$$\le \frac{1}{2} \|x\|_{-1}^2 + \delta \int_0^t \int_{\mathscr{O}} |\eta_\lambda(s)| d\xi ds - \lambda \int_0^t \int_{\mathscr{O}} y_\lambda(s)(y_\lambda(s) + W_G(s)) d\xi ds.$$

Since

$$-y_\lambda(s)(y_\lambda(s) + W_G(s)) \le -\frac{1}{2} \|y_\lambda(s)\|^2 + \frac{1}{2} W_G^2(s),$$

we find that

$$\frac{1}{2} \|y_\lambda(t)\|_{-1}^2 + \int_0^t \int_{\mathscr{O}} (j(z_\lambda(s)) + j^*(\eta_\lambda(s)) d\xi ds + \frac{\lambda}{2} \int_0^t \int_{\mathscr{O}} |y_\lambda(s)|^2 d\xi ds$$

$$+ \frac{1}{2\lambda} \int_0^t \int_{\mathscr{O}} (y_\lambda(s) + W_G(s) - z_\lambda(s))^2 d\xi ds$$

$$\le \left(\frac{1}{2} \|x\|_{-1}^2 + \delta \int_0^t \int_{\mathscr{O}} |\eta_\lambda(s)| d\xi ds + \frac{\lambda}{2} \int_0^t \int_{\mathscr{O}} W_G^2(s) d\xi ds \right), \quad t \in [0, T]. \tag{5.31}$$

On the other hand, we recall (see (5.5)) that condition $D(\beta) = \mathbb{R}$ is equivalent with

$$\lim_{|p| \to \infty} \frac{j^*(p)}{|p|} = +\infty. \tag{5.32}$$

So, there exists $N = N(\omega)$ such that

$$|\eta_\lambda(s)| > N \Rightarrow j^*(\eta_\lambda(s)) > 2C\delta|\eta_\lambda(s)|.$$

Consequently, for $C > |Q_T|$ we have that

$$\int_0^t \int_{\mathscr{O}} |\eta_\lambda(s)| d\xi ds = \int \int_{|\eta_\lambda(s)| > N} |\eta_\lambda(s)| d\xi ds + \int \int_{|\eta_\lambda(s)| \leq N} |\eta_\lambda(s)| d\xi ds$$
$$\leq \frac{1}{2C\delta} \int_0^t \int_{\mathscr{O}} j^*(\eta_\lambda(s)) d\xi ds + NC\delta.$$

Substituting this into (5.31), since $j \geq 0$, we obtain (5.26), which in particular implies

$$\int_0^t \int_{\mathscr{O}} (j(z_\lambda(s)) + j^*(\eta_\lambda(s))) d\xi ds \leq C_1(1 + \|x\|_{-1}^2), \tag{5.33}$$

and

$$\int_0^t \int_{\mathscr{O}} (y_\lambda + W_G - z_\lambda)^2 d\xi ds \leq 2\lambda C_1(1 + \|x\|_{-1}^2). \tag{5.34}$$

By (5.26) it follows that $\{y_\lambda\}$ is bounded and therefore weak-star compact in $L^\infty(0, T; H^{-1})$

5.2.2 Convergence for $\lambda \to 0$

Since

$$\lim_{|u| \to \infty} j(u)/|u| = \infty, \quad \lim_{|u| \to \infty} j^*(u)/|u| = \infty, \tag{5.35}$$

we deduce from (5.33) that the sequences $\{z_\lambda\}$ and $\{\eta_\lambda\}$ are bounded and equi-integrable in $L^1(Q_T)$. Then by the Dunford-Pettis theorem (Theorem 1.2.12) the sequences $\{z_\lambda\}$ and $\{\eta_\lambda\}$ are weakly compact in $L^1(Q_T)$. Hence along a subsequence, again denoted by λ, we have

$$z_\lambda \to z, \quad \eta_\lambda \to \eta \quad \text{weakly in } L^1(Q_T) \text{ as } \lambda \to 0. \tag{5.36}$$

Moreover, by (5.34) we see that $z = y + W_G$, where

$$y_\lambda \to y \quad \text{weak-star in } L^\infty(0, T; H^{-1}) \text{ and weakly in } L^1(Q_T). \tag{5.37}$$

Also note that by (5.22) we have for every $t \in [0, T]$

$$y_\lambda(t) - \Delta\left(\int_0^t (\eta_\lambda(s) + \lambda(y_\lambda(s) + W_G(s))) ds\right) = x \tag{5.38}$$

and so the sequence $\{\int_0^\bullet (\eta_\lambda(s) + \lambda y_\lambda(s))ds\}$ is bounded in $L^\infty(0, T; H_0^1)$. Hence, selecting a further subsequence if necessary (see (5.28)), we have

$$\lim_{\lambda \to 0} \int_0^\bullet (\eta_\lambda(s) + \lambda y_\lambda(s))ds = \int_0^\bullet \eta(s)ds \quad \text{weakly* in } L^\infty(0, T; H_0^1). \qquad (5.39)$$

So, by (5.38) we find

$$y(t) - \Delta \int_0^t \eta(s)ds = x, \quad \text{a.e. } t \in [0, T]. \qquad (5.40)$$

Since

$$\int_0^\bullet \eta(s)ds \in C([0, T]; L^1) \cap L^\infty(0, T; H_0^1),$$

$t \mapsto \int_0^t \eta(s)ds$ is weakly continuous in H_0^1 and therefore we infer that so is $t \mapsto -\Delta \int_0^t \eta(s)ds$ in H^{-1}. Thus the function

$$\tilde{y}(t) := -\Delta \int_0^t \eta(s)ds + x, \quad t \in [0, T], \qquad (5.41)$$

is an H^{-1}-valued weakly continuous version of y. Furthermore, we claim that for $\lambda \to 0$

$$y_\lambda(t) \to \tilde{y}(t) \quad \text{weakly in } H^{-1}, \quad \forall t \in [0, T].$$

Indeed, since $\eta_\lambda \to \eta$ weakly in $L^1(Q_T)$ and $\lambda(y_\lambda + W_G) \to 0$ weakly in $L^1(Q_T)$ (since it even converges strongly in $L^2(Q_T)$ to zero by (5.30)), it follows that for every $t \in [0, T]$

$$\int_0^t (\eta_\lambda(s) + \lambda(y_\lambda(s) + W_G(s)))ds \to \int_0^t \eta(s)ds \quad \text{weakly in } L^1.$$

Hence by (5.38) and the definition of $\tilde{\eta}$ we obtain that for every $t \in [0, T]$

$$(-\Delta)^{-1} y_\lambda(t) \to (-\Delta)^{-1} \tilde{y}(t) \quad \text{weakly in } L^1.$$

Since by (5.31) $y_\lambda(t)$, $\lambda > 0$, are bounded in H^{-1}, the above immediately implies the claim.

From now on we shall consider this particular version $\tilde{y}$ of y defined in (5.41) and for simplicity we denote it again by y; so we have

$$y_\lambda(t) \to y(t) \quad \text{weakly in } H^{-1}, \quad \forall t \in [0, T].$$

We can also rewrite Eq. (5.41) as

$$y_t(t) - \Delta\eta(t) = 0 \quad \text{in } \mathscr{D}'(Q_T), \quad y(0) = x. \tag{5.42}$$

Now we are going to show that

$$\eta(t,\xi) \in \beta(y(t,\xi) + W_G(t,\xi)) \quad \text{a.e. } (t,\xi) \in Q_T. \tag{5.43}$$

For this we shall need the following inequality

$$\liminf_{\lambda\to 0} \int_{Q_T} y_\lambda \eta_\lambda d\xi dt \le \int_{Q_T} y\eta d\xi dt. \tag{5.44}$$

We first recall Eq. (5.27) which yields

$$j_\lambda(y_\lambda + W_G) + j_\lambda^*(\eta_\lambda) = (y_\lambda + W_G)\eta_\lambda, \quad \text{a.e. in } Q_T,$$

and so by (5.9), we have

$$j((1 + \lambda\beta)^{-1}(y_\lambda + W_G)) + j^*(\eta_\lambda) \le (y_\lambda + W_G)\eta_\lambda \quad \text{a.e. in } Q_T,$$

because $j_\lambda^*(p) \ge j_\lambda(p), \ \forall \ \lambda > 0, \ p \in \mathbb{R}$. This yields

$$\int_{Q_T} (j(z_\lambda) + j^*(\eta_\lambda))d\xi dt \le \int_{Q_T} (y_\lambda + W_G)\eta_\lambda d\xi dt.$$

Since the convex functional

$$(z,\zeta) \to \int_{Q_T} (j(z) + j^*(\zeta))d\xi dt$$

is lower semicontinuous on $L^1(Q_T) \times L^1(Q_T)$ (and consequently weakly lower semicontinuous on this space) we obtain by (5.36) that

$$\int_{Q_T} (j(y + W_G) + j^*(\eta))d\xi dt \le \liminf_{\lambda\to 0} \int_{Q_T} y_\lambda \eta_\lambda d\xi dt + \int_{Q_T} W_G \eta d\xi dt. \tag{5.45}$$

Furthermore, by (5.28), (5.36) and again by the weak lower semicontinuity of convex integrals in $L^1(Q_T)$ it follows that

$$j(y + W_G), \ j^*(\eta) \in L^1(Q_T). \tag{5.46}$$

On the other hand, since $j(u) + j^*(p) \ge up$ for all $u, p \in \mathbb{R}$, we have

$$(W_G + y)\eta \le j(y + W_G) + j^*(\eta) \quad \text{a. e. in } Q_T. \tag{5.47}$$

Moreover, by assumption (5.2) we see that for every $M > 0$ there exists $R = R(M) \geq 0$, such that

$$j(-y - W_G) \leq Mj(y + W_G) \quad \text{on } Q^R$$

where

$$Q^R = \{(t, \xi) \in Q_T : \ |y(t, \xi) + W_G(t, \xi)| \geq R\}.$$

Since $j(y + W_G) \in L^1(Q_T)$ we have, by continuity of j,

$$j(-y - W_G) \leq h \quad \text{a. e. in } Q_T, \tag{5.48}$$

where $h \in L^1(Q_T)$. On the other hand, since j is bounded from below we have

$$j(-y - W_G) \in L^1(Q_T). \tag{5.49}$$

Taking into account that by virtue of the same inequality (5.4), besides (5.47), we have that

$$-(y + W_G)\eta \leq j(-y - W_G) + j^*(\eta) \quad \text{a. e. in } Q_T, \tag{5.50}$$

by (5.47) and (5.48) it follows that a. e. in Q_T we have

$$|(W_G + y)\eta| \leq \max\{j(y + W_G) + j^*(\eta), \ j(-y - W_G) + j^*(\eta)\} \in L^1(Q_T)$$

and therefore in particular $y\eta \in L^1(Q_T)$ (recall that $W_G \in L^\infty(Q_T)$).

Now we come back to Eq. (5.24) which by integration yields

$$\frac{1}{2} \left(\|y_\lambda(T)\|_{-1}^2 - \|x\|_{-1}^2 \right) + \int_{Q_T} y_\lambda \eta_\lambda d\xi dt + \lambda \int_{Q_T} y_\lambda(y_\lambda + W_G) d\xi dt = 0. \tag{5.51}$$

Taking into account that

$$y_\lambda(T) \to y(T) \quad \text{weakly in } H, \tag{5.52}$$

by (5.51) we obtain

$$\liminf_{\lambda \to 0} \int_{Q_T} y_\lambda \eta_\lambda d\xi dt \leq -\frac{1}{2} \left(\|y(T)\|_{-1}^2 - \|x\|_{-1}^2 \right). \tag{5.53}$$

In order to complete the proof one needs an integration by parts formula in Eq. (5.41) (or (5.42)), obtained multiplying the equation by y and integrating on Q_T. Formally this is possible because $y\eta \in L^1(Q_T)$ and $y(t) \in H^{-1}$ for all $t \in [0, T]$. But, in order to prove it rigorously, one must give sense to $(y'(t), y(t))$. Lemma 5.2.1 below answers this question positively and by (5.53) also proves (5.44).

We first note that since j, j^* are nonnegative and convex and $j(0) = 0 = j^*(0)$, we have for all measurable $f : Q_T \to \mathbb{R}$ and $\alpha \in [0, 1]$,

$$j(f) \in L^1(Q_T) \Rightarrow j(\alpha f) \in L^1(Q_T).$$

Since by convexity

$$j(\alpha y) = j(\alpha(y + W_G) - \alpha W_G) \le \alpha j(y + W_G) + (1 - \alpha) j(-\tfrac{\alpha W_G}{1-\alpha})$$

and $W_G \in L^\infty(Q_T)$ by (5.47) we have $j(\alpha y) \in L^1(Q_T)$. Moreover, since $j^*(\eta) \in L^1(Q_T)$, we have $j^*(\alpha \eta) \in L^1(Q_T)$. Hence, y and η constructed above fulfill all conditions in the following lemma.

Lemma 5.2.1 *Let $y \in C^w([0, T]; H^{-1})$ and $\eta \in L^1(Q_T)$ satisfy*

$$y(t) - \Delta \int_0^t \eta(s)ds = x, \quad t \in [0, T]. \tag{5.54}$$

Furthermore, assume that for some $\alpha > 0$, $j(\alpha y)$, $j(-\alpha y)$, $j^*(\alpha \eta) \in L^1(Q_T)$ and $\eta \in L^1(Q_T)$. Then $y\eta \in L^1(Q_T)$ and

$$\int_{Q_T} y\eta \, d\xi dt = -\frac{1}{2} \left(\|y(T)\|_{-1}^2 - \|x\|_{-1}^2 \right) \tag{5.55}$$

$$Y_\varepsilon \Sigma_\varepsilon \to y\eta \quad in \ L^1(Q_T), \ for \ \epsilon \to 0,$$

where Y_ε, Σ_ε are defined in (5.56) below.

Proof We set for $\varepsilon > 0$, and $A = -\Delta$, $D(A) = H^2 \cap H_0^1$,

$$Y_\varepsilon = (1 + \varepsilon A)^{-m} y, \quad \Sigma_\varepsilon = (1 + \varepsilon A)^{-m} \eta, \tag{5.56}$$

where $m \in \mathbb{N}$ is such that $m > \max\{2, (d+2)/4\}$. Then by elliptic regularity,

$$Y_\varepsilon \in C^w([0, T]; H_0^1 \cap H^{2m-1}) \subset C^w([0, T]; H_0^1) \cap C(\overline{\mathcal{O}}))$$

and

$$\Sigma_\varepsilon \in L^1(0, T; W^{2,q}), \quad 1 < q < \frac{d}{d-1}.$$

Hence $Y_\varepsilon \Sigma_\varepsilon \in L^1(Q_T)$ and for $\varepsilon \to 0$

$$\begin{cases} Y_\varepsilon(t) \to y(t) & \text{strongly in } H^{-1}, \ \forall\, t \in [0,T] \\ Y_\varepsilon \to y & \text{strongly in } L^1(Q_T) \\ \Sigma_\varepsilon \to \eta & \text{strongly in } L^1(Q_T) \\ \int_0^t \Sigma_\varepsilon(s)ds \to \int_0^t \eta(s)ds & \text{strongly in } H_0^1, \ \forall\, t \in [0,T]. \end{cases} \tag{5.57}$$

Here we note that the last fact follows because (5.54) implies that $\int_0^\bullet \eta(s)ds \in C^w([0,T]; H_0^1)$. We have also by (5.54)

$$Y_\varepsilon(t) + A \int_0^t \Sigma_\varepsilon(s)ds = (1 + \varepsilon A)^{-m}x, \quad \forall\, t \in [0,T],$$

which implies

$$\frac{d}{dt} Y_\varepsilon(t) + A\Sigma_\varepsilon(t) = 0, \quad t \in [0,T]$$

and, taking inner product in H^{-1} with $Y_\varepsilon(t)$, we obtain

$$\frac{1}{2} \frac{d}{dt} \|Y_\varepsilon(t)\|_{-1}^2 + \int_{\mathcal{O}} \Sigma_\varepsilon(t) Y_\varepsilon(t)d\xi = 0, \quad \text{a.e. } t \in [0,T].$$

Hence

$$\lim_{\varepsilon \to 0} \int_{Q_T} \Sigma_\varepsilon(t) Y_\varepsilon(t) d\xi\, dt = -\frac{1}{2} \left(\|y(T)\|_{-1}^2 - \|x\|_{-1}^2 \right) \tag{5.58}$$

and by (5.57) we may assume that for $\varepsilon \to 0$

$$Y_\varepsilon \to y, \quad \Sigma_\varepsilon \to \eta \quad \text{a. e. in } Q_T. \tag{5.59}$$

Moreover, by (5.4) we have

$$\alpha^2 \Sigma_\varepsilon Y_\varepsilon \le j(\alpha Y_\varepsilon) + j^*(\alpha \Sigma_\varepsilon), \quad -\alpha^2 \Sigma_\varepsilon Y_\varepsilon \le j(-\alpha Y_\varepsilon) + j^*(\alpha \Sigma_\varepsilon) \quad \text{a. e. in } Q_T. \tag{5.60}$$

Now we claim that for $\varepsilon \to 0$

$$j(\alpha Y_\varepsilon) \to j(\alpha y), \ j^*(\alpha \Sigma_\varepsilon) \to j^*(\alpha \eta), \ j(-\alpha Y_\varepsilon) \to j(-\alpha y) \quad \text{in } L^1(Q_T). \tag{5.61}$$

By (5.59) these convergences hold a.e. in Q_T. So, in order to prove (5.61) it suffices to show that $\{j(\alpha Y_\varepsilon)\}, \{j^*(\alpha \Sigma_\varepsilon)\}, \{j(-\alpha Y_\varepsilon)\}$ are equi-integrable on Q_T and so, again by the Dunford–Pettis Theorem, that they are weakly compact in $L^1(Q_T)$. To this end let $y \in L^1$ and let $Y_\varepsilon := (1 + \varepsilon A)^{-1}y$, i.e. Y_ε is the solution to the Dirichlet problem

$$\begin{cases} Y_\varepsilon - \varepsilon \Delta Y_\varepsilon = y, & \text{in } \mathcal{O}, \\ \\ Y_\varepsilon = 0, & \text{on } \partial\mathcal{O}. \end{cases} \tag{5.62}$$

It may be represented as

$$Y_\varepsilon(\xi) = \int_{\mathcal{O}} G(\xi, \xi_1) y(\xi_1) d\xi_1, \quad \forall\, \xi \in \mathcal{O}, \tag{5.63}$$

where G is the associated Green function. It is well known that $\int_{\mathcal{O}} G(\xi, \xi_1) d\xi_1$ is the solution to (5.62) with $y = 1$ so that by the maximum principle we have $0 < \int_{\mathcal{O}} G(\xi, \xi_1) d\xi_1 \leq 1$ for all $\xi \in \mathcal{O}$.

We may rewrite Y_ε as

$$Y_\varepsilon(\xi) = \int_{\mathcal{O}} G(\xi, \xi_2) d\xi_2 \int_{\mathcal{O}} \tilde{G}(\xi, \xi_1) y(\xi_1) d\xi_1, \quad \forall\, \xi \in \mathcal{O},$$

where

$$\tilde{G}(\xi, \xi_1) = \frac{G(\xi, \xi_1)}{\int_{\mathcal{O}} G(\xi, \xi_2) d\xi_2}$$

and so $\int_{\mathcal{O}} \tilde{G}(\xi, \xi_1) d\xi_1 = 1$ for all $\xi \in \mathcal{O}$.

Then, if $j(y) \in L^1, j(0) = 0$ by Jensen's inequality we have

$$j(Y_\varepsilon(\xi)) \leq \int_{\mathcal{O}} G(\xi, \xi_2) d\xi_2 \int_{\mathcal{O}} \tilde{G}(\xi, \xi_1) j(y(\xi_1)) d\xi_1$$

$$= \int_{\mathcal{O}} G(\xi, \xi_1) j(y(\xi_1)) d\xi_1, \quad \forall\, \xi \in \mathcal{O}.$$

Hence we have shown that for any $y \in L^1$ with $j(y) \in L^1$,

$$j((1 + \varepsilon A)^{-1}y) \leq (1 + \varepsilon A)^{-1} j(y), \quad \forall\, \xi \in \mathcal{O}.$$

Iterating and using the fact that $(1 + \varepsilon A)^{-1}$ preserves positivity we get for all $m \in \mathbb{N}$

$$j((1 + \varepsilon A)^{-m}y) \leq (1 + \varepsilon A)^{-m} j(y), \quad \text{a.e. in } \mathcal{O}. \tag{5.64}$$

Now let y be as in the assertion of the lemma and Y_ε as in (5.56). Integrating over Q_T, since $(1 + \varepsilon A)^{-m}$ is a contraction on L^1, (5.64) applied to αy implies

$$\int_{Q_T} j(\alpha Y_\varepsilon(\xi, t)) d\xi dt \le \int_{Q_T} j(\alpha y(\xi_2, t)) d\xi_2 dt.$$

Taking into account that $j(\alpha y) \in L^1(Q_T)$ we infer that $\{j(\alpha Y_\varepsilon)\}$ is equi-integrable on Q_T. The same argument applies to $\{j^*(\alpha \Sigma_\varepsilon)\}$, $\{j(-\alpha Y_\varepsilon)\}$.

Then (5.60) implies that the sequence $\{\Sigma_\varepsilon Y_\varepsilon\}$ is equi-integrable on Q_T and consequently by the Dunford-Pettis theorem, weakly compact in $L^1(Q_T)$. Since $\{\Sigma_\varepsilon Y_\varepsilon\}$ is a.e. convergent to $y\eta$ we infer that for $\varepsilon \to 0$

$$\Sigma_\varepsilon Y_\varepsilon \to y\eta \quad \text{strongly in } L^1(Q_T), \tag{5.65}$$

which combined with (5.58) implies (5.55) as desired.

5.2.2.1 Proof of (5.43)

We have

$$j(z_\lambda) - j(u) \le \eta_\lambda(z_\lambda - u), \quad \forall u \in \mathbb{R} \quad \text{a.e. in } Q_T.$$

Integrating over Q_T yields

$$\int_{Q_T} j(z_\lambda) d\xi dt \le \int_{Q_T} j(u) d\xi dt + \int_{Q_T} \eta_\lambda(z_\lambda - u) d\xi dt, \quad \forall u \in L^\infty(Q_T).$$

Note that by the definition of β_λ we have

$$z_\lambda = -\lambda \eta_\lambda + y_\lambda + W_G.$$

Therefore, since $z = y + W_G$, by (5.44) and Fatou's lemma we can let $\lambda \to 0$ to obtain

$$\int_{Q_T} j(z) d\xi dt - \int_{Q_T} j(u) d\xi dt \le \int_{Q_T} \eta(z - u) d\xi dt, \quad \forall u \in L^\infty(Q_T).$$

Now by Lusin's theorem for each $\epsilon > 0$ there is a compact subset $Q_\epsilon \subset Q_T$ such that $(d\xi \otimes dt)(Q_T \setminus Q_\epsilon) \le \epsilon$ and y, η are continuous on Q_ϵ. Let (t_0, x_0) be a Lebesgue point for y, η and $y\eta$ and let B_r be the ball of center (t_0, x_0) and radius r. We take

$$u(t, \xi) = \begin{cases} z(t, \xi), & \text{if } (t, \xi) \in Q_\epsilon \cap B_r^c \\ v, & \text{if } (t, \xi) \in (Q_\epsilon \cap B_r) \cup (Q_T \setminus Q_\epsilon). \end{cases}$$

Here v is arbitrary in $\mathbb{R}$. Since u is bounded we can substitute it into the above inequality to get

$$\int_{B_r \cap Q_\epsilon} (j(z) - j(v) - \eta(z - v))d\xi dt \leq \int_{Q_T \setminus Q_\epsilon} (\eta(z - v) + j(v) - j(z))d\xi dt.$$

Letting $\epsilon \to 0$ we obtain that

$$\int_{B_r} (j(z) - j(v) - \eta(z - v))d\xi dt \leq 0, \quad \forall v \in \mathbb{R}, \ r > 0.$$

This yields

$$j(z(t_0, x_0)) \leq j(v) + \eta(t_0, x_0)(z(t_0, x_0) - v), \quad \forall v \in \mathbb{R}.$$

and therefore $\eta(t_0, x_0) \in \partial j(z(t_0, x_0)) = \beta(z(t_0, x_0))$. Since almost all points of Q_T are Lebesgue points we get (5.43) as claimed.

5.2.3 Completion of Proof of Theorem 5.1.4

Let us first summarize that we have proved the existence of a pair $(y, \eta) \in L^1(Q_T) \times L^1(Q_T)$, satisfying

$$
\begin{aligned}
& y \in C^w([0, T]; H^{-1}), \quad \int_0^\bullet \eta(s)ds \in C^w([0, T]; H_0^1), \\
& \eta(t, \xi) \in \beta(y(t, \xi)) \quad \text{for a.e. } (t, \xi) \in Q^T, \\
& y(t) - \Delta \int_0^t \eta(s)ds = x, \quad t \in [0, T], \\
& j(\alpha y), \ j^*(\alpha y) \in L^1(Q_T) \quad \text{for } \alpha \in (0, 1].
\end{aligned}
$$

We claim that (y, η) is the only such pair. Indeed, if $(\tilde{y}, \tilde{\eta})$ is another one then

$$j(\tfrac{\alpha}{2}(y - \tilde{y})) \leq \tfrac{1}{2} j(\alpha y) + \tfrac{1}{2} j(-\alpha \tilde{y})$$

and

$$j^*(\tfrac{\alpha}{2}(y - \tilde{y})) \leq \tfrac{1}{2} j^*(\alpha y) + \tfrac{1}{2} j^*(-\alpha \tilde{y}).$$

But as we have seen before Lemma 5.2.1 the right hand sides are in $L^1(Q_T)$. Hence $y - \tilde{y}, \eta - \tilde{\eta}$ fulfill all conditions of Lemma 5.2.1 and adopting the notation above we have for $\varepsilon > 0$

$$
\begin{aligned}
Y_\varepsilon(t) - \tilde{Y}_\varepsilon(t) &= \Delta \int_0^t (\Sigma_\varepsilon(s) - \tilde{\Sigma}_\varepsilon(s))ds \\
&= \int_0^t \Delta(\Sigma_\varepsilon(s) - \tilde{\Sigma}_\varepsilon(s))ds, \quad t \in [0, T].
\end{aligned}
$$

Differentiating and subsequently taking the inner product in H^{-1} with $Y_\varepsilon(t) - \tilde{Y}_\varepsilon(t)$ and integrating again we arrive at

$$\frac{1}{2} \left\| (1 + \varepsilon A)^{-m} (Y_\varepsilon(t) - \tilde{Y}_\varepsilon(t)) \right\|_{-1}^2 = \int_0^t \int_{\mathscr{O}} (Y_\varepsilon(s) - \tilde{Y}_\varepsilon(s)) (\Sigma_\varepsilon(s) - \tilde{\Sigma}_\varepsilon(s)) d\xi ds$$

$$= \int_0^t \int_{\mathscr{O}} ((1 + \varepsilon A)^{-m} (y(s) - \tilde{y}(s)) (1 + \varepsilon A)^{-m} (\eta(s) - \tilde{\eta}(s)) d\xi ds, \quad t \in [0, T].$$

Letting $\varepsilon \to 0$ and applying Lemma 5.2.1 we obtain that for $t \in [0, T]$

$$\frac{1}{2} \|y(t) - \tilde{y}(t)\|_{-1}^2 = \int_0^t \int_{\mathscr{O}} (y(s) - \tilde{y}(s)) (\eta(s) - \tilde{\eta}(s)) d\xi ds \leq 0$$

by the monotonicity of β. This proves the uniqueness of (y, η).

Now let us consider the ω-dependence of y and η. By (5.41), (5.43) we know that $y = y(t, \xi, \omega)$ is the solution to the equation

$$\begin{cases} y'(t) - \Delta\beta(y(t) + W_G(t)(\omega)) = 0 & \text{a.e. } t \in [0, T], \\ y(0) = x \end{cases} \tag{5.66}$$

and as seen earlier for $\eta = \eta(t, \xi, \omega)$ as in (5.36)

$$y \in C^w([0, T]; H^{-1}) \cap L^1(Q_T), \quad \eta \in L^1(Q_T)$$

$$\int_0^{\bullet} \eta(s) ds \in C^w([0, T]; H_0^1), \tag{5.67}$$

$$\eta(t, \xi, \omega) \in \beta(y(t, \xi, \omega)) + W_G(t, \xi, \omega) \quad \text{a.e. } (t, \xi, \omega) \in Q_T \times \Omega. \tag{5.68}$$

By the above uniqueness of (y, η), it follows that for any sequence $\lambda \to \infty$ we have $\mathbb{P}$-a.s.

$$y_\lambda(t) \to y(t) \quad \text{weakly in } H^{-1}, \ \forall t \in [0, T],$$

$$y_\lambda \to y \quad \text{weakly in } L^1(Q_T),$$

$$\int_0^t \eta_\lambda(s) ds \to \int_0^t \eta(s) ds \quad \text{weakly in } L^1, \ \forall t \in [0, T]$$

$$\text{and weakly in } H_0^1, \text{ a.e. } t \in [0, T],$$

$$\eta_\lambda \to \eta \quad \text{weakly in } L^1(Q_T).$$

Therefore y and η are strong $L^1(Q_T)$-limits of a sequence of convex combinations of y_λ, η_λ respectively, and since y_λ and η_λ are predictable processes, it follows that so are y and η. In particular, this means that $Y(t) = y(t) + W_G(t)$ is an H^{-1}-valued weakly continuous adapted process and that the following equation is satisfied

$$Y(t) - \Delta \int_0^t \eta(s)ds = x + \int_0^t G(s)dW(s), \quad t \in [0, T]. \tag{5.69}$$

Equivalently

$$\begin{cases} dY(t) - \Delta\beta(Y(t))dt = G(t)dW(t), \\ \\ Y(0) = x. \end{cases} \tag{5.70}$$

In order to prove that Y is a solution of (5.70) in the sense of Definition 5.1.2 with $G(t)$ replacing $B(X(t))$, uniqueness and some energy estimates we need an Itô's formula type result. As in the case of Lemma 5.2.1 the difficulty is that the integral

$$\int_{Q_T} \beta(Y)Y d\xi dt$$

(if β is multi-valued by $\beta(Y)$ we mean a L^1 section of $\beta(Y)$) might be (in general) not well defined taking into account that $\beta(Y), Y \in L^1(Q_T)$ only. We have, however

Lemma 5.2.2 *Let Y be a H^{-1}-valued weakly continuous adapted process satisfying equation* (5.69). *Then the following equality holds* $\mathbb{P}$-*a.s.*

$$\frac{1}{2}\|Y(t)\|_{-1}^2 = \frac{1}{2}\|x\|_{-1}^2 - \int_0^t \int_{\mathscr{O}} \eta(s)Y(s)d\xi ds$$

$$\tag{5.71}$$

$$+ \int_0^t \langle Y(s), G(s)dW(s)\rangle_{-1} + \frac{1}{2}\int_0^t \|G(s)\|_{L_{HS}(L^2, H^{-1})}^2 ds, \quad \mathbb{P}\text{-}a.s.$$

Furthermore, $Y \in C_W([0, T]; H^{-1}) \cap L^1((0, T) \times \mathscr{O} \times \Omega)$, *and* $\eta \in L^1((0, T) \times \mathscr{O} \times \Omega)$ *and all conditions* (5.10)–(5.13) *are satisfied.*

Proof By Lemma 5.2.1 we have that $Y\eta \in L^1(Q_T)$. Next we introduce the sequences (see the proof of Lemma 5.2.1)) for $m \in \mathbb{N}$

$$Y_\varepsilon = (1 + \varepsilon A)^{-m}Y, \quad \Sigma_\varepsilon = (1 + \varepsilon A)^{-m}\eta.$$

For large enough m we can apply Itô's formula to the equation

$$\begin{cases} dY_\varepsilon(t) + A\Sigma_\varepsilon(t) = (1 + \varepsilon A)^{-m}GdW(t) \\ \\ Y_\varepsilon(0) = (1 + \varepsilon A)^{-m}x = x_\varepsilon, \end{cases} \tag{5.72}$$

and obtain

$$\frac{1}{2}\,\|Y_\varepsilon(t)\|_{-1}^2 = \frac{1}{2}\,\|x_\varepsilon\|_{-1}^2 - \int_0^t \int_{\mathcal{O}} \Sigma_\varepsilon(s) Y_\varepsilon(s) d\xi ds$$

$$+ \int_0^t \langle Y_\varepsilon(s), G_\varepsilon(s) dW(s)\rangle_{-1} + \frac{1}{2} \int_0^t \|G_\varepsilon(s)\|_{LHS(L^2(\mathcal{O}),H)}^2 ds, \quad t \in [0,T].$$
(5.73)

where $G_\varepsilon = (1 + \varepsilon A)^{-m} G$. Letting $\varepsilon \to 0$ (since $W_G \in L^\infty(Q_T)$) we get by (5.65)

$$\int_{Q_T} Y_\varepsilon \Sigma_\varepsilon d\xi ds \to \int_{Q_T} Y \eta d\xi ds, \quad \mathbb{P}\text{-a.s.}.$$

Furthermore

$$Y_\varepsilon(t) \to Y(t) \quad \text{strongly in } H^{-1}, \ \forall\, t \in [0,T],$$

which by virtue of (5.73) yields (5.71), if we one proves first that for $t \in [0,T]$

$$\mathbb{P} - \lim_{\varepsilon \to 0} \int_0^t \langle Y_\varepsilon(s), G_\varepsilon(s) dW(s)\rangle = \int_0^t \langle Y(s), G(s) dW(s)\rangle.$$
(5.74)

We shall even show that this convergence in probability is locally uniform in t. We have by a standard consequence of the Burkholder-Davis-Gundy inequality for $p = 1$ (see 1.2.2) that for $\bar{Y}_\varepsilon := (1 + \varepsilon A)^{-2m} Y$ and $\delta_1, \delta_2 > 0$

$$\mathbb{P}\left[\sup_{t \in [0,T]} \left| \int_0^t \langle Y(s), G(s) dW(s)\rangle - \int_0^t \langle Y_\varepsilon(s), G_\varepsilon(s) dW(s)\rangle \right| \geq \delta_1 \right]$$
(5.75)

$$\leq \frac{3\delta_2}{\delta_1} + \mathbb{P}\left[\int_0^T \|G(s)\|_{LHS(L^2, H^{-1})}^2 |Y(s) - \bar{Y}_\varepsilon(s)|_{-1}^2 ds \geq \delta_2 \right].$$

Since $Y \in C^w([0,T]; H^{-1})$, $\mathbb{P}$-a.s. and $(1 + \varepsilon A)^{-1}$ is a contraction on H^{-1} we have

$$\sup_{s \in [0,T]} \|Y(s) - \bar{Y}_\varepsilon(s)\|_{-1} \leq 2 \sup_{s \in [0,T]} \|Y(s)\|_{-1}^2, \quad \mathbb{P}\text{-a.s.}.$$

Hence by (5.18) the second term on the right hand side of (5.75) converges to zero as $\varepsilon \to 0$. Taking subsequently $\delta_2 \to 0$, (5.75) implies (5.74). We emphasize that, since the left hand size of (5.71) is not continuous $\mathbb{P}$-a.s. (though all terms on the right hand side are), the $\mathbb{P}$-zero set of $\omega \in \Omega$ for which (5.71) does not hold might depend on t.

Next we want to prove that

$$\mathbb{E} \sup_{t \in [0,T]} \|Y(t)\|_{-1}^2 < \infty. \tag{5.76}$$

To this end first note that by (5.68) and (5.3) we have

$$\eta(s)Y(s) = j(Y(s)) + j^*(\eta(s)) \geq 0, \tag{5.77}$$

hence (5.71) implies that for every $t \in [0, T]$

$$\|Y(t)\|_{-1}^2 \leq \|x\|_{-1}^2 + N_t + \int_0^t \|G(s)\|_{\mathscr{L}_2(L^2, H^{-1})}^2 ds, \quad \mathbb{P}\text{-a.s.}, \tag{5.78}$$

where

$$N_t := \int_0^t \langle Y(s), G(s) dW(s) \rangle_{-1}, \quad t \geq 0,$$

is a continuous local martingale such that

$$< N >_t = 2 \int_0^t |G^*(s)Y(s)|_{L^2}^2 ds, \quad t \geq 0,$$

where $G^*(s)$ is the adjoint operator of $G(s) : L^2 \to H^{-1}$. We shall prove that

$$\mathbb{E} \sup_{t \in [0,T]} |N_t| < +\infty. \tag{5.79}$$

By the Burkholder-Davis-Gundy inequality (Proposition 1.2.2) for $p = 1$ applied to stopping times

$$\tau_N := \inf\{t \geq 0 : |N_t| \geq N\} \wedge T, \quad N \in \mathbb{N},$$

we obtain

$$\mathbb{E}\left[\sup_{t \in [0, \tau_N]} |N_t| \right] \leq 3\mathbb{E}\left[\sup_{s \in [0, \tau_N]} \|Y(s)\|_{-1} \left(4 \int_0^{\tau_N} \|G(s)\|_{\mathscr{L}(L^2; H^{-1})}^2 ds \right)^{1/2} \right]$$

$$\leq 6C \left(\mathbb{E}\left[\sup_{s \in [0, \tau_N]} \|Y(s)\|_{-1}^2 \right] \right)^{1/2}, \tag{5.80}$$

where

$$C := \left(\mathbb{E}\left[\int_0^T \|G(s)\|_{\mathscr{L}(L^2; H^{-1})}^2 ds \right] \right)^{1/2} < \infty.$$

Since $Y \in C^w([0,T]; H^{-1})$, we know that $s \mapsto \|Y(s)\|_{-1}^2$ is weakly lower semicontinuous. Therefore by (5.78)

$$\sup_{s \in [0,\tau_N]} \|Y(s)\|_{-1}^2 = \sup_{s \in [0,\tau_N] \cap \mathbb{Q}} \|Y(s)\|_{-1}^2 \le \|x\|_{-1}^2 + \sup_{s \in [0,\tau_N]} |N_s|$$

$$+ \int_0^T \|G(s)\|_{\mathscr{L}_2(L^2; H^{-1})}^2 ds, \quad \mathbb{P}\text{-a.s.}.$$

So (5.80) implies that for all $N \in \mathbb{N}$

$$\left(\mathbb{E} \left[\sup_{t \in [0,\tau_N]} |N_t| \right] \right)^2 \le 36C^2 \left[\|x\|_{-1}^2 + \mathbb{E} \left[\sup_{s \in [0,\tau_N]} |N_s| \right] + C^2 \right],$$

which entails that

$$\sup_{N \in \mathbb{N}} \mathbb{E} \sup_{t \in [0,\tau_N]} |N_t| < \infty.$$

By monotone convergence this implies (5.79), since N_t has continuous sample paths and $\tau_N \uparrow T$ as $N \to \infty$. Now (5.78) implies that (5.14) holds.

By (5.79) and (5.71) it follows that

$$\eta Y \in L^1((0,T) \times \mathscr{O} \times \Omega). \tag{5.81}$$

Hence we infer as above that

$$j(Y), j^*(\eta) \in L^1((0,T) \times \mathscr{O} \times \Omega)$$

and therefore

$$Y, \eta \in L^1((0,T) \times \mathscr{O} \times \Omega).$$

Taking expectation in (5.71) we see that $t \mapsto \mathbb{E}[\|Y(t)\|_{-1}^2]$ is continuous. Since $Y \in C^w([0,T]; H^{-1})$, $\mathbb{P}$-a.s., (5.81) then also implies that $Y \in C_W([0,T]; H^{-1})$. This in turn together with (5.69) implies that also holds.

Now we come back to the proof of Theorem 5.1.4. We first note that Lemma 5.2.2 also implies the uniqueness of the solution Y and estimate (5.19). Indeed by (5.75) and monotonicity of β we have for YG_i, $i = 1, 2$ the estimate (5.19). This concludes the proof of Theorem 5.1.4.

5.2.4   *Proof of Theorem 5.1.4*

Consider the space

$$\mathscr{K} = \Big\{ X \in C_W([0,T]; H^{-1}) \cap L^1((0,T) \times \mathcal{O} \times \Omega) : X \text{ predictable,}$$

$$\sup_{t \in [0,T]} \mathbb{E}[e^{-2\alpha t} \|X(t)\|_{-1}^2] \le M_1^2, \ \mathbb{E} \int_{Q_T} j(X(s)) d\xi ds \le M_2 \Big\},$$

where $\alpha > 0, M_1 > 0$ and $M_2 > 0$ will be specified later.

The space $\mathscr{K}$ is endowed with the metric induced by the norm

$$\|X\|_\alpha = \left(\sup_{t \in [0,T]} \mathbb{E}[e^{-2\alpha t} \|X(t)\|_{-1}^2] \right)^{1/2}.$$

Note that $\mathscr{K}$ is closed in the norm $\| \cdot \|_\alpha$. Indeed, if $X_n \to X$ in $\| \cdot \|_\alpha$ then since

$$\mathbb{E} \int_{Q_T} j(X_n(s)) d\xi ds \le M_2, \quad \forall \, n \in \mathbb{N},$$

(5.35) implies that

$$X_n \to X, \quad \text{in } L^1((0,T) \times \mathcal{O} \times \Omega)$$

and by Fatou's Lemma we get

$$\mathbb{E} \int_{Q_T} j(X(s)) d\xi ds \le M_2.$$

as claimed. Now consider the mapping $\Gamma : \mathscr{K} \to C_W([0,T]; H^{-1}) \cap L^1((0,T) \times \mathcal{O} \times \Omega)$ defined by

$$Y = \Gamma(X),$$

where $Y \in C_W([0,T]; H^{-1}) \cap L^1((0,T) \times \mathcal{O} \times \Omega)$ is the solution in the sense of Definition 5.1.2 of the problem

$$\begin{cases} dY(t) - \Delta \beta(Y(t)) dt = B(X(t)) dW(t) & \text{in } Q_T, \\[2mm] \beta(Y(t)) = 0 & \text{on } \Sigma_T, \\[2mm] Y(0) = x & \text{in } \mathcal{O}. \end{cases}$$

We shall prove that for α, M_1, M_2 suitably chosen, Γ maps $\mathscr{K}$ into itself and it is a contraction in the norm $\| \cdot \|_\alpha$.

Indeed by (5.71) and (5.3) for any solution Y to (5.23) we have that

$$\frac{1}{2} \|Y(t)\|_{-1}^2 + \int_0^t \int_{\mathscr{O}} (j(Y(s)) + j^*(\eta(s))) d\xi ds$$

$$= \int_0^t \langle Y(s), B(X(s))dW(s) \rangle_{-1}$$

$$+ \frac{1}{2} \int_0^t \|B(X(s))\|_{L_{HS}(L^2;H^{-1})}^2 ds + \frac{1}{2} \|x\|_{-1}^2, \quad t \in [0,T], \ \mathbb{P}\text{-a.s.}.$$

By Hypothesis 7(ii) we have

$$\frac{1}{2} \sup_{t \in [0,T]} \mathbb{E}[e^{-2\alpha t} \|Y(t)\|_{-1}^2] + e^{-2\alpha t} \mathbb{E} \int_0^t \int_{\mathscr{O}} (j(Y(s)) + j^*(\eta(s))) d\xi ds$$

$$\leq \frac{1}{2} \|x\|_{-1}^2 + \frac{L^2}{2} \sup_{t \in [0,T]} \left[e^{-2\alpha t} \int_0^t \mathbb{E}|X(s)|_{-1}^2 ds \right]$$

$$\leq \frac{1}{2} \|x\|_{-1}^2 + \frac{L^2}{2} \sup_{t \in [0,T]} \int_0^t e^{-2\alpha(t-s)} \mathbb{E} e^{-2\alpha s} \|X(s)\|_{-1}^2 ds \leq \frac{1}{2} \|x\|_{-1}^2 + \frac{L^2 M_1^2}{4\alpha}.$$

Hence

$$\sup_{t \in [0,T]} \mathbb{E}[e^{-2\alpha t} \|Y(t)\|_{-1}^2] \leq \frac{L^2 M_1^2}{2\alpha} + \|x\|_{-1}^2$$

and

$$\mathbb{E} \int_{Q_T} (j(Y(s)) + j^*(\eta(s)))) d\xi \leq \left(\frac{L^2 M_1^2}{2\alpha} + |x|_{-1}^2 \right) e^{2\alpha T}.$$

Hence for $\alpha > L^2$, $M_1^2 > 2\|x\|_{-1}^2$ and $M_2 \geq M_1^2 e^{2\alpha T}$ we have that $Y \in \mathscr{K}$ and the operator Γ maps $\mathscr{K}$ into itself. By a similar computation involving Hypothesis 7(ii) we see that for M_1, M_2 and α suitably chosen we have

$$\|Y_1 - Y_2\|_\alpha \leq \frac{C}{\sqrt{\alpha}} \|X_1 - X_2\|_\alpha$$

where $Y_i = \Gamma X_i$, $i = 1, 2$. Hence for a suitable α, Γ is a contraction and so equation $X = \Gamma(X)$ has a unique solution in Γ. This completes the proof.

$\square$

Remark 5.2.3 Theorem 5.1.4 can be closely compared with Theorem 3.4.1. It is established, however, under weaker assumptions on the growth of β at $\pm\infty$. For instance a function β of the form

$$\beta(r) = ae^{\alpha|r|} \operatorname{sign} r, \quad \forall\, r \in \mathbb{R},$$

where $a, \alpha > 0$ is covered by Theorem 5.1.4 but not by the existence theory in Chap. 3 which besides monotonicity required polynomial growth of β.

5.3   Comments and Bibliographical Remarks

The main result of this chapter, Theorem 5.1.4 was established in [22] and the proof given here closely follows this paper. A similar result on stochastic porous media equation driven by Lèvy noise was obtained in [13]. These results can be compared most closely with that obtained in Chap. 4 or that in [84] via variational arguments in Orlicz spaces. There is not, however, a large overlap and the methods are completely different.

Chapter 6
The Stochastic Porous Media Equations in $\mathbb{R}^d$

Here we shall treat Eq. (3.1) in the domain $\mathscr{O} = \mathbb{R}^d$. Though the methods are similar to those used for bounded domains, there are, however, some notable differences and as seen below the dimension d of the space plays a crucial role.

It should be mentioned also that in the deterministic case the existence to a weak solution X to equation

$$\begin{cases} \dfrac{\partial X}{\partial t} - \Delta \beta(X) = 0, & \text{in } \mathscr{D}'((0,T) \times \mathbb{R}^d), \\[2mm] X(0) = x, & \text{in } \mathbb{R}^d, \end{cases} \tag{6.1}$$

follows by the Crandall–Liggett theorem (see e.g. [6, p. 233]). But the stochastic equation considered here is more delicate and apparently no transfer from the deterministic case is possible.

6.1 Introduction

We consider here Eq. (2.1) in the domain $\mathscr{O} = \mathbb{R}^d$ and for $\sigma(X) = X$. Namely,

$$\begin{cases} dX - \Delta \beta(X) dt = X dW, & \text{in } (0,T) \times \mathbb{R}^d, \\[2mm] X(0) = x, & \text{in } \mathbb{R}^d, \end{cases} \tag{6.2}$$

© Springer International Publishing Switzerland 2016
V. Barbu et al., *Stochastic Porous Media Equations*, Lecture Notes
in Mathematics 2163, DOI 10.1007/978-3-319-41069-2_6

where β is a monotonically nondecreasing function on $\mathbb{R}$ (possibly multivalued) and $W(t)$ is a Wiener process of the form

$$W(t) = \sum_{k=1}^{\infty} \mu_k e_k W_k(t), \ t \geq 0. \tag{6.3}$$

Here $\{W_k\}$ are independent Brownian motions on a filtered probability space $\{\Omega, \mathscr{F}, \mathscr{F}_t, \mathbb{P}\}$, $\mu_k \in \mathbb{R}$ and $\{e_k\}$ is an orthonormal basis in $H^{-1}(\mathbb{R}^d)$ or $\mathscr{H}^{-1}$ (see (6.6) below) to be made precise later on.

On bounded domains $\mathscr{O} \subset \mathbb{R}^d$ with Dirichlet homogeneous boundary conditions, Eq. (6.2) was studied in previous chapters under various assumptions on β. It should be said, however, that there is a principal difference between bounded and unbounded domains, mainly due to the multiplier problem in Sobolev spaces on $\mathbb{R}^d$.

We study the existence and uniqueness of (6.2) under two different sets of conditions requiring a different functional approach. The first one, which will be presented in Sect. 6.3, assumes that β is monotonically nondecreasing and Lipschitz. The state space for (6.2) is, in this case, $H^{-1}(\mathbb{R}^d)$, that is, the dual of the classical Sobolev space $H^1(\mathbb{R}^d)$.

The second case, which will be studied in Sect. 6.4, is the one where β is a maximal monotone multivalued function with polynomial growth. An important physical problem covered by this case is the self-organized criticality model (met in Chap. 3 on a bounded set)

$$dX - \Delta H(X - X_c)dt = (X - X_c)dW, \tag{6.4}$$

where H is the Heaviside function and X_c is the critical state (see Sect. 1.1).

We note that in this second case, the solution $X(t)$ to (6.2) is defined in a certain distribution space $\mathscr{H}^{-1}$ (see (6.3) below) on $\mathbb{R}^d$ and the existence is obtained for $d \geq 3$ only. The case $1 \leq d \leq 2$ remains open due to the absence of a multiplier rule in the norm $\| \cdot \|_{\mathscr{H}^{-1}}$ (see Lemma 6.4.1 below).

Finally, in Sect. 6.5 we prove the finite time extinction of the solution X to (6.2) with strictly positive probability under the assumption that $\beta(r)r \geq \rho|r|^{m+1}$ and $m = \frac{d-2}{d+2}$.

6.2 Preliminaries

To begin with, let us briefly recall a few definitions pertaining distribution spaces on $\mathbb{R}^d$, whose classical Euclidean norm will be denoted by $|\cdot|$.

Denote by $\mathscr{S}(\mathbb{R}^d)$ the space of rapidly decreasing functions on $\mathbb{R}^d$ and $\mathscr{S}'$ its dual, that is the space of all temperate distributions on $\mathbb{R}^d$ (see, e.g., [67]). Denote by $\mathscr{H}$ the space

$$\mathscr{H} = \{\varphi \in \mathscr{S}'/\mathbb{R} : \xi \mapsto |\xi|\widetilde{\mathscr{F}}(\varphi)(\xi) \in L^2(\mathbb{R}^d)\}, \tag{6.5}$$

where $\mathscr{S}'/\mathbb{R}$ is the factor space of $\mathscr{S}'$ modulo $\mathbb{R}$ and $\widetilde{\mathscr{F}}$ is the Fourier transform of φ. In other words,

$$\mathscr{S}'/\mathbb{R} = \{\varphi = \psi + C : \psi \in \mathscr{S}', \ C \in \mathbb{R}\}$$

and $\varphi = \widetilde{\varphi}$ if $\varphi - \widetilde{\varphi} \equiv$ constant.

$$\widetilde{\mathscr{F}}(\psi)(x) := \int_{\mathbb{R}^d} \exp\{-ix \cdot \xi\} \psi(\xi) \, d\xi, \quad \forall x \in \mathbb{R}^d.$$

We denote by $L^2(\mathbb{R}^d)$ the space of square integrable functions on $\mathbb{R}^d$ with norm $|\cdot|_2$ and scalar product $\langle \cdot, \cdot \rangle_2$. In general $|\cdot|_p$ will denote the norm of $L^p(\mathbb{R}^d)$ or $L^p(\mathbb{R}^d; \mathbb{R}^d)$, $1 \le p \le \infty$. The dual space $\mathscr{H}^{-1}$ of $\mathscr{H}$ is given by

$$\mathscr{H}^{-1} = \{\eta \in \mathscr{S}' : \xi \mapsto \mathscr{F}(\eta)(\xi)|\xi|^{-1} \in L^2(\mathbb{R}^d)\}. \tag{6.6}$$

As seen by (6.6), the space $\mathscr{H}^{-1}$ can be equivalently defined as

$$\{\eta \in \mathscr{S}' : \eta = \nabla u, \quad u \in L^2(\mathbb{R}^d)\}.$$

The duality between $\mathscr{H}$ and $\mathscr{H}^{-1}$ is denoted by $\langle \cdot, \cdot \rangle$ and is given by

$$\langle \varphi, \eta \rangle = \int_{\mathbb{R}^d} \widetilde{\mathscr{F}}(\varphi)(\xi)\overline{\mathscr{F}(\eta)(\xi)}d\xi \tag{6.7}$$

and the norm of $\mathscr{H}$ denoted by $|\cdot|_1$ is given by

$$|\varphi|_1 = \left(\int_{\mathbb{R}^d} |\widetilde{\mathscr{F}}(\varphi)(\xi)|^2 |\xi|^2 d\xi\right)^{\frac{1}{2}} = \left(\int_{\mathbb{R}^d} |\nabla \varphi|^2 d\xi\right)^{\frac{1}{2}}. \tag{6.8}$$

(We note that (6.8) is a norm for the factor space $\mathscr{H}$, that is $|\varphi|_1 = 0 \Rightarrow \varphi = C$.)
   The norm of $\mathscr{H}^{-1}$, denoted by $|\cdot|_{-1}$ is given by

$$|\eta|_{-1} = \left(\int_{\mathbb{R}^d} |\xi|^{-2} |\widetilde{\mathscr{F}}(\eta)(\xi)|^2 d\xi\right)^{\frac{1}{2}} = [\langle (-\Delta)^{-1}\eta, \eta \rangle]^{\frac{1}{2}}. \tag{6.9}$$

(We note that the operator $-\Delta$ is an isomorphism from $\mathscr{H}$ onto $\mathscr{H}^{-1}$.) The scalar product of $\mathscr{H}^{-1}$ is given by

$$\langle \eta_1, \eta_2 \rangle_{-1} = \langle (-\Delta)^{-1}\eta_1, \eta_2 \rangle. \tag{6.10}$$

As regards the relationship of $\mathscr{H}$ with the factor space $L^p(\mathbb{R}^d)/\mathbb{R}$, we have the following.

Lemma 6.2.1 *Let $d \geq 3$. Then*

$$\mathscr{H} \subset L^{\frac{2d}{d-2}}(\mathbb{R}^d)/\mathbb{R} \tag{6.11}$$

with continuous and dense injection.

Indeed, by the Sobolev embedding theorem (see, e.g., [36, p. 278]), we have

$$|\varphi|_{\frac{2d}{d-2}} \leq C|\nabla\varphi|_2, \quad \forall \varphi \in C^\infty(\mathbb{R}^d),$$

and this implies (6.11), as claimed.

It should be mentioned that (6.11) is no longer true for $1 \leq d \leq 2$. However, by duality, we have via the Hahn–Banach theorem

$$L^{\frac{2d}{d+2}}(\mathbb{R}^d) \subset \mathscr{H}^{-1}, \quad \forall \, d \geq 3. \tag{6.12}$$

We recall that H^1 is the Sobolev space

$$H^1 = \{u \in L^2(\mathbb{R}^d) : \nabla u \in L^2(\mathbb{R}^d)\}$$

$$= \{u \in L^2(\mathbb{R}^d) : \xi \mapsto \widetilde{\mathscr{F}}(u)(\xi)(1 + |\xi|^2)^{1/2} \in L^2(\mathbb{R}^d)\},$$

(it should be emphasized here that $\|\varphi\|_1$ is a factor norm, that is $\|\varphi\|_1 = \inf\{|\nabla\varphi|_2 + |\alpha|, \ \alpha \in \mathbb{R}\}$) with norm

$$|u|_{H^1} = \left((u^2 + |\nabla u|^2)d\xi\right)^{\frac{1}{2}} = \left(|\mathscr{F}(u)(\xi)(1 + |\xi|^2)^{1/2}\right)^{\frac{1}{2}}$$

and by H^{-1} its dual, that is,

$$H^{-1} = \{u \in \mathscr{S}'(\mathbb{R}^d) : \mathscr{F}(u)(\xi)(1 + |\xi|^2)^{-\frac{1}{2}} \in L^2(\mathbb{R}^d)\}.$$

The norm of H^{-1} is denoted by $\|\cdot\|_{-1}$ and its scalar product by $\langle\cdot,\cdot\rangle_{-1}$. We have the continuous and dense embeddings

$$H^1(\mathbb{R}^d) \subset \mathscr{H}, \quad \mathscr{H}^{-1} \subset H^{-1}.$$

It should be emphasized, however, that $\mathscr{H}$ is not a subspace of $L^2(\mathbb{R}^d)$ and so $L^2(\mathbb{R}^d)$ is not the pivot space in the duality $\langle\cdot,\cdot\rangle$ given by (6.7).

6.3 Equation (6.2) with a Lipschitzian β

Consider here Eq. (6.2) under the following conditions.

Hypothesis 8

(i) β : ℝ → ℝ is monotonically nondecreasing, Lipschitz such that β(0) = 0.
(ii) W is a Wiener process as in (6.3), where $e_k \in H^1(\mathbb{R}^d)$, such that

$$C_\infty^2 := 36 \sum_{k=1}^\infty \mu_k^2 (|\nabla e_k|_\infty^2 + |e_k|_\infty^2 + 1) < \infty, \tag{6.13}$$

and $\{e_k\}$ is an orthonormal basis in $H^{-1}(\mathbb{R}^d)$.

As for equations in bounded domains the Lipschitz case is first step toward more general maximal nonlinearities β. On the other hand, it has an intrinsic interest.

Definition 6.3.1 Let $x \in H^{-1}(\mathbb{R}^d)$. A continuous, $(\mathscr{F}_t)_{t \geq 0}$-adapted process X : $[0, T] \to H^{-1}(\mathbb{R}^d)$ is called a *distributional solution* to (6.2) if the following conditions hold

(i) $X \in L^2(\Omega; C([0, T]; H^{-1}(\mathbb{R}^d))) \cap L^2([0, T] \times \Omega; L^2(\mathbb{R}^d))$.

(ii) $\int_0^\bullet \beta(X(s))ds \in C([0, T]; H^1(\mathbb{R}^d))$,   ℙ-a.e.

(iii) We have

$$X(t) - \Delta \int_0^t \beta(X(s))ds = x + \int_0^t X(s)dW(s), \quad \forall\, t \in [0, T],\ \mathbb{P}\text{-a.e.} \tag{6.14}$$

Remark 6.3.2 As in Sect. 1.2, the stochastic Itô's integral in (6.14) is the standard one from [51] or [82]. In fact, in the terminology of these references, W is a Q-Wiener process W^Q on H^{-1}, where $Q : H^{-1} \to H^{-1}$ is the symmetric trace class operator defined by

$$Qh := \sum_{k=1}^\infty \mu_k^2 \langle e_k, h \rangle_{-1} e_k, \quad h \in H^{-1}.$$

For $x \in H^{-1}$, define $\sigma(x) : Q^{1/2}H^{-1} \to H^{-1}$ by

$$\sigma(x)(Q^{1/2}h) = \sum_{k=1}^\infty (\mu_k \langle e_k, h \rangle_{-1} e_k \cdot x), \quad h \in H. \tag{6.15}$$

By (6.13), each e_k is an H^{-1}-multiplier such that

$$|e_k x|_{-1} \leq 2 (|e_k|_\infty + |\nabla e_k|_\infty) |x|_{-1}, \quad x \in H^{-1}. \tag{6.16}$$

Hence, for all $x \in H^{-1}$,

$$\sum_{k=1}^{\infty} |\mu_k \langle e_k, h \rangle_{-1} e_k x|_{-1} \leq \left(\sum_{k=1}^{\infty} \mu_k^2 |e_k x|_{-1}^2 \right)^{1/2} |h|_{-1}$$

$$\leq 2C_\infty |x|_{-1} |h|_{-1} = 2C_\infty |x|_{-1} |Q^{1/2} h|_{Q^{1/2} H^{-1}},$$

and thus $\sigma(x)$ is well-defined and an element in $L(Q^{1/2} H^{-1}, H^{-1})$. Moreover, for $x \in H^{-1}$, by (6.15), (6.16),

$$\|\sigma(x)\|_{L_2(Q^{1/2} H^{-1}, H^{-1})}^2 = \sum_{k=1}^{\infty} |\sigma(x)(Q^{1/2} e_k)|_{-1}^2 = \sum_{k=1}^{\infty} |\mu_k e_k x|_{-1}^2$$

$$= \sum_{k=1}^{\infty} \mu_k^2 |e_k x|_{-1}^2 \leq C_\infty^2 |x|_{-1}^2. \tag{6.17}$$

Since $\{ Q^{1/2} e_k \mid k \in \mathbb{N} \}$ is an orthonormal basis of $Q^{1/2} H^{-1}$, it follows that $\sigma(x) \in \mathscr{L}_2(Q^{1/2} H^{-1}, H^{-1})$ and the map $x \mapsto \sigma(x)$ is linear and continuous (hence Lipschitz) from H^{-1} to $\mathscr{L}_2(Q^{1/2} H^{-1}, H^{-1})$. Hence (e.g., according to [82, Sect. 2.3])

$$\int_0^t X(s) dW(s) := \int_0^t \sigma(X(s)) dW^Q(s), \quad t \in [0, T],$$

is well-defined as a continuous H^{-1}-valued martingale and by Itô's isometry and (6.17)

$$\mathbb{E} \left\| \int_0^t X(s) dW(s) \right\|_{-1}^2 = \sum_{k=1}^{\infty} \mu_k^2 \mathbb{E} \int_0^t \|X(s) e_k\|_{-1}^2 ds$$

$$\leq C_\infty^2 \mathbb{E} \int_0^t \|X(s)\|_{-1}^2 ds, \quad t \in [0, T]. \tag{6.18}$$

Furthermore, it follows that

$$\int_0^t X(s) dW(s) = \sum_{k=1}^{\infty} \int_0^t \sigma(X(s))(Q^{1/2} e_k) dW_k(s)$$

$$= \sum_{k=1}^{\infty} \int_0^t \mu_k e_k X(s) dW_k(s), \quad t \in [0, T], \tag{6.19}$$

where the series converges in $L^2(\Omega; C([0, T]; H^{-1}))$.

In fact, $\int_0^\bullet X(s)dW(s)$ is a continuous L^2-valued martingale, because $X \in L^2([0, T] \times \Omega; L^2(\mathbb{R}^d))$ and, analogously to (6.17), we get

$$\|\sigma(x)\|^2_{\mathscr{L}_2(Q^{1/2}H^{-1},L^2)} \leq C_\infty^2 |x|_2^2, \quad x \in L^2(\mathbb{R}^d).$$

In particular, by Itô's isometry,

$$\mathbb{E}\left|\int_0^t X(s)dW(s)\right|_2^2 \leq C_\infty^2 \mathbb{E}\int_0^t |X(s)|_2^2 ds, \quad t \in [0, T].$$

Furthermore, the series in (6.18) is also convergent in $L^2(\Omega; C([0, T]; L^2(\mathbb{R}^d)))$. □

Theorem 6.3.3 *Let $d \geq 1$ and $x \in L^2(\mathbb{R}^d)$. Then, under Hypothesis 8(i)(ii) there is a unique strong solution to Eq. (6.2). This solution satisfies*

$$\mathbb{E}\left[\sup_{t\in[0,T]} |X(t)|_2^2\right] \leq 2|x|_2^2 e^{3C_\infty^2 t}.$$

In particular, $X \in L^2(\Omega; L^\infty([0, T]; L^2(\mathbb{R}^d)))$. Assume further that

$$\beta(r)r \geq \alpha r^2, \quad \forall r \in \mathbb{R}, \tag{6.20}$$

where $\alpha > 0$. Then, there is a unique strong solution X to (6.2) for all $x \in H^{-1}(\mathbb{R}^d)$.

Proof We approximate (6.2) by

$$\begin{cases} dX + (v - \Delta)\beta(X)dt = XdW(t), & t \in (0, T), \\ X(0) = x, & \text{on } \mathbb{R}^d, \end{cases} \tag{6.21}$$

where $v \in (0, 1)$. Then we first we prove Lemma 6.3.4 below.

Lemma 6.3.4 *Assume that β is as in Hypothesis 8(i) and let $x \in L^2(\mathbb{R}^d)$. Then, there is a unique $(\mathscr{F}_t)_{t\geq 0}$-adapted solution $X = X^v$ to (6.21) in the following strong sense:*

$$X^v \in L^2(\Omega, C([0, T]; H^{-1}(\mathbb{R}^d))) \cap L^2([0, T] \times \Omega; L^2(\mathbb{R}^d)), \tag{6.22}$$

and $\mathbb{P}$-a.e.

$$X^v(t) = x + (\Delta - v)\int_0^t \beta(X^v(s))ds + \int_0^t X^v(s)dW(s), \ t \in [0, T]. \tag{6.23}$$

In addition, for all $v \in (0, 1)$,

$$\mathbb{E}\left[\sup_{t \in [0,T]} |X^v(t)|_2^2\right] \leq 2|x|_2^2 e^{3C_\infty^2 T}. \tag{6.24}$$

If, moreover, β *satisfies* (6.20), *then for each* $x \in H^{-1}(\mathbb{R}^d)$ *there is a unique solution* X^v *satisfying* (6.22) *and* (6.23).

Proof Let us start with the second part of the assertion, i.e., we assume that β satisfies (6.20) and that $x \in H^{-1}(\mathbb{R}^d)$. Then the standard theory developed in Chap. 4 (see also [82, Sects. 4.1 and 4.2]) applies to ensure that there exists a unique solution X^v taking value in $H^{-1}(\mathbb{R}^d)$ satisfying (6.22) and (6.23) above. Indeed, it is easy to check that (H1)–(H4) from [82, Sect. 4.1] are satisfied with $V := L^2(\mathbb{R}^d)$, $H := H^{-1}(\mathbb{R}^d)$, $Au := (\Delta - v)(\beta(u))$, $u \in V$, and $H^{-1}(\mathbb{R}^d)$ is equipped with the equivalent norm

$$\|\eta\|_{-1,v} := \langle \eta, (v - \Delta)H^{-1}\eta\rangle^{1/2}, \ \eta \in H^{-1}(\mathbb{R}^d)$$

(in which case, we also write H_v^{-1}). Here, as before, we use $\langle \cdot, \cdot \rangle$ also to denote the dualization between $H^{-1}(\mathbb{R}^d)$ and $H^{-1}(\mathbb{R}^d)$. For details, we refer to the calculations in [82, Example 4.1.11], which since $p = 2$, go through when the bounded domain Ω there is replaced by $\mathbb{R}^d$. Hence [82, Theorem 4.2.4] applies to give the above solution X^v.

In the case when β does not satisfy (6.20), the above conditions (H1), (H2), (H4) from [82] still hold, but (H3) not in general. Therefore, we replace β by $\beta + \lambda I$, $\lambda \in (0, 1)$, and thus consider $A_\lambda(u) := (\Delta - v)(\beta(u) + \lambda u)$, $u \in V := L^2(\mathbb{R}^d)$ and, as above, by [82, Theorem 4.2.4], obtain a solution X_λ^v, satisfying (6.22), (6.23), to

$$\begin{cases} dX_\lambda^v(t) + (v - \Delta)(\beta(X_\lambda^v(t)) + \lambda X_\lambda^v(t))dt = X_\lambda^v(t)dW(t), \ t \in [0, T], \\ X_\lambda^v(0) = x \in H^{-1}(\mathbb{R}^d). \end{cases} \tag{6.25}$$

In particular, by (6.22),

$$\mathbb{E}\left[\sup_{t \in [0,T]} \|X_\lambda^v(t)\|_{-1}^2\right] < \infty. \tag{6.26}$$

We want to let $\lambda \to 0$ to obtain a solution to (6.21). To this end, in this case (i.e., without assuming (6.20)), we assume from now on that $x \in L^2(\mathbb{R}^d)$. The reason is that we need the following:

Claim 1 We have $X_\lambda^\nu \in L^2([0,T] \times \Omega; H^1(\mathbb{R}^d))$ and

$$\mathbb{E}\left[\sup_{t \in [0,T]} |X_\lambda^\nu(t)|_2^2\right] + 4\lambda\mathbb{E}\int_0^T |\nabla X_\lambda^\nu(s)|_2^2 ds \leq 2|x|_2^2 e^{3C_\infty^2 T},$$

for all $\nu, \lambda \in (0,1)$. Furthermore, X_λ^ν has continuous sample paths in $L^2(\mathbb{R}^d)$, $\mathbb{P}$-a.e.

To prove the claim we note that

$$X_\lambda^\nu(t) = x + (\Delta - \nu)\int_0^t (\beta(X_\lambda^\nu(s)) + \lambda X_\lambda^\nu(s))ds$$

$$+ \int_0^t X_\lambda^\nu(s)dW(s), \ t \in [0,T]. \tag{6.27}$$

Let $\alpha \in (\nu, \infty)$. Recalling that $(\alpha - \Delta)^{-\frac{1}{2}} : H^{-1}(\mathbb{R}^d) \to L^2(\mathbb{R}^d)$ and applying this operator to the above equation, we find

$$(\alpha - \Delta)^{-\frac{1}{2}}X_\lambda^\nu(t) = (\alpha - \Delta)^{-\frac{1}{2}}x + \int_0^t (\Delta - \nu)(\alpha - \Delta)^{-\frac{1}{2}}(\beta(X_\lambda^\nu(s)) + \lambda X_\lambda^\nu(s))ds$$

$$+ \int_0^t (\alpha - \Delta)^{-\frac{1}{2}}\sigma(X_\lambda^\nu(s))Q^{1/2}dW(s), \ t \in [0,T]. \tag{6.28}$$

Applying Itô's formula with $H = L^2(\mathbb{R}^d))$ to $|(\alpha - \Delta)^{-\frac{1}{2}}X_\lambda^\nu(t)|_2^2$, we obtain, for $t \in [0,T]$,

$$|(\alpha - \Delta)^{-\frac{1}{2}}X_\lambda^\nu(t)|_2^2 = |(\alpha - \Delta)^{-\frac{1}{2}}x|_2^2$$

$$+2\int_0^t \langle(\Delta - \nu)(\alpha - \Delta)^{-\frac{1}{2}}\beta(X_\lambda^\nu(s)), (\alpha - \Delta)^{-\frac{1}{2}}X_\lambda^\nu(s)\rangle ds$$

$$-2\lambda\int_0^t (|\nabla((\alpha - \Delta)^{-\frac{1}{2}}X_\lambda^\nu(s))|_2^2 + \nu|(\alpha - \Delta)^{-\frac{1}{2}}X_\lambda^\nu(s)|_2^2)ds$$

$$+\int_0^t \|(\alpha - \Delta)^{-\frac{1}{2}}\sigma(X_\lambda^\nu(s))Q^{1/2}\|_{L_2(H^{-1},L^2)}^2 ds$$

$$+2\int_0^t \langle(\alpha - \Delta)^{-\frac{1}{2}}X_\lambda^\nu(s), (\alpha - \Delta)^{-\frac{1}{2}}\sigma(X_\lambda^\nu(s))Q^{1/2}dW(s)\rangle_2. \tag{6.29}$$

But, for $f \in L^2(\mathbb{R}^d)$, we have

$$(\alpha - \Delta)^{-\frac{1}{2}}(\Delta - \nu)(\alpha - \Delta)^{-\frac{1}{2}}f = (P - I)f,$$

where

$$P := (\alpha - \nu)(\alpha - \Delta)^{-1}.$$

For the Green function g_α of $(\alpha - \Delta)$, we then have, for $f \in L^2(\mathbb{R}^d)$,

$$Pf = (\alpha - \nu) \int_{\mathbb{R}^d} f(\xi)g_\alpha(\cdot, \xi)d\xi.$$

Hence, by [86, Lemma 5.1], the integrand of the second term on the right hand side of (6.29) with $f := X_\lambda^\nu(s)$ ($\in L^2(\mathbb{R}^d)$ for ds-a.e. $s \in [0, T]$) can be rewritten as

$$\langle \beta(f), (P - I)f \rangle_2 = -\frac{1}{2} \int_{\mathbb{R}^d} \int_{\mathbb{R}^d} [\beta(f(\bar{\xi})) - \beta(f(\xi))][f(\bar{\xi}) - f(\xi)]g_\alpha(\xi, \bar{\xi})d\bar{\xi}\,d\xi$$

$$- \int_{\mathbb{R}^d} (1 - P1(\xi)) \cdot \beta(f(\xi))f(\xi)d\xi.$$

Since β is monotone, $\beta(0) = 0$ and $P1 \leq 1$, we deduce that

$$\langle \beta(f), (P - I)f \rangle \leq 0.$$

Hence, after a multiplication by α, (6.29) implies that, for all $t \in [0, T]$ (see Remark 6.3.2),

$$\alpha|(\alpha - \Delta)^{-\frac{1}{2}}X_\lambda^\nu(t)|_2^2 + 2\lambda \int_0^t |\nabla(\sqrt{\alpha}(\alpha - \Delta)^{-\frac{1}{2}}X_\lambda^\nu(s))|_2^2 ds$$

$$\leq \alpha|(\alpha - \Delta)^{-\frac{1}{2}}x|_2^2 + \int_0^t \sum_{k=1}^\infty \mu_k^2 \langle \alpha(\alpha - \Delta)^{-1}(e_k X_\lambda^\nu(s)), e_k X_\lambda^\nu(s) \rangle_2 ds$$

$$+ 2 \int_0^t \langle \alpha(\alpha - \Delta)^{-1}X_\lambda^\nu(s), \sigma(X_\lambda^\nu(s))Q^{1/2}dW(s) \rangle_2.$$

Hence, by the Burkholder–Davis–Gundy inequality, (1.23) (with $p = 1$) since $\alpha(\alpha - \Delta)^{-1}$ is a contraction on $L^2(\mathbb{R}^d)$,

$$
\mathbb{E}\left[\sup_{s\in[0,t]} |\sqrt{\alpha}(\alpha - \Delta)^{-\frac{1}{2}} X_\lambda^\nu(s)|_2^2\right] + 2\lambda \mathbb{E}\int_0^t |\nabla(\sqrt{\alpha}(\alpha - \Delta)^{-\frac{1}{2}} X_\lambda^\nu(s))|_2^2 ds
$$
$$
\le |\sqrt{\alpha}(\alpha - \Delta)^{-\frac{1}{2}} x|_2^2 + C_\infty^2 \mathbb{E}\int_0^t |X_\lambda^\nu(s)|_2^2 ds
$$
$$
+ 6\mathbb{E}\left(\int_0^t \sum_{k=1}^\infty \mu_k^2 \langle \alpha(\alpha - \Delta)^{-1} X_\lambda^\nu(s), e_k X_\lambda^\nu(s)\rangle_2^2 ds\right)^{1/2}.
$$

$$(6.30)$$

The latter term can be estimated by

$$
C_\infty \mathbb{E}\left[\sup_{s\in[0,t]} |\alpha(\alpha - \Delta)^{-1} X_\lambda^\nu(s)|_2 \left(\int_0^t |X_\lambda^\nu(s)|_2^2 ds\right)^{1/2}\right]
$$
$$
\le \frac{1}{2}\mathbb{E}\left\{\sup_{s\in[0,t]} |\sqrt{\alpha}(\alpha - \Delta)^{-\frac{1}{2}} X_\lambda^\nu(s)|_2^2\right\} + \frac{1}{2}C_\infty^2 \mathbb{E}\int_0^t |X_\lambda^\nu(s)|_2^2 ds,
$$

$$(6.31)$$

where we used that $\sqrt{\alpha}(\alpha - \Delta)^{-\frac{1}{2}}$ is a contraction on $L^2(\mathbb{R}^d)$. Note that the first summand on the right hand side is finite by (6.26), since the norm $|\sqrt{\alpha}(\alpha - \Delta)^{-\frac{1}{2}} \cdot|_2$ is equivalent to $\|\cdot\|_{-1}$. Hence, we can subtract this term after substituting (6.31) into (6.30) to obtain

$$
\mathbb{E}\left[\sup_{s\in[0,t]} |\sqrt{\alpha}(\alpha - \Delta)^{-\frac{1}{2}} X_\lambda^\nu(s)|_2^2\right] + 4\lambda \mathbb{E}\int_0^t |\nabla(\sqrt{\alpha}(\alpha - \Delta)^{-\frac{1}{2}} X_\lambda^\nu(s)|_2^2 ds
$$
$$
\le 2|\sqrt{\alpha}(\alpha - \Delta)^{-\frac{1}{2}} x|_2^2 + 3C_\infty^2 \mathbb{E}\int_0^t |X_\lambda^\nu(s)|_2^2 ds, \quad t \in [0, T].
$$

$$(6.32)$$

Obviously, the quantity under the supremum on the left hand side of (6.32) is increasing in α. So, by the monotone convergence theorem, we may let $\alpha \to \infty$ in (6.32) and then, except for its last part, Claim 1 immediately follows by Gronwall's lemma, since $\sqrt{\alpha}(\alpha - \Delta)^{-\frac{1}{2}}$ is a contraction in $L^2(\mathbb{R}^d)$ and $x \in L^2(\mathbb{R}^d)$. The last part of Claim 1 then immediately follows from [71, Theorem 2.1].

Applying Itô's formula to $\|X_\lambda^\nu(t) - X_{\lambda'}^\nu(t)\|_{-1,\nu}^2$ it follows from (6.27) that, for $\lambda, \lambda' \in (0, 1)$ and $t \in [0, T]$,

$$
\|X_\lambda^\nu(t) - X_{\lambda'}^\nu(t)\|_{-1,\nu}^2 + 2\int_0^t \langle \beta(X_\lambda^\nu) - \beta(X_{\lambda'}^\nu) + (\lambda X_\lambda^\nu - \lambda' X_{\lambda'}^\nu), X_\lambda^\nu - X_{\lambda'}^\nu\rangle_2 ds
$$
$$
= \int_0^t \|\sigma(X_\lambda^\nu(s)) - \sigma(X_{\lambda'}^\nu(s))\|_{\mathscr{L}_2(Q^{1/2}H^{-1}, H_\nu^{-1})}^2 ds
$$
$$
+ 2\int_0^t \langle X_\lambda^\nu(s) - X_{\lambda'}^\nu(s), \sigma(X_\lambda^\nu(s) - X_{\lambda'}^\nu(s)) dW^Q(s)\rangle_{-1,\nu}.
$$

$$(6.33)$$

Our Hypothesis 8(i) on β implies that

$$(\beta(r) - \beta(r'))(r - r') \geq (Lip\,\beta + 1)^{-1}|\beta(r) - \beta(r')|^2 \quad \text{for } r, r' \in \mathbb{R},$$

where $Lip\,\beta$ is the Lipschitz constant of β. Hence (6.33), (6.17) and the inequality (1.23) (for $p = 1$) imply that, for all $t \in [0, T]$,

$$\mathbb{E}\left[\sup_{s\in[0,t]} \|X_\lambda^\nu(s) - X_{\lambda'}^\nu(s)\|_{-1,\nu}^2\right] + 2(Lip\,\beta + 1)^{-1}\mathbb{E}\int_0^t |\beta(X_\lambda^\nu(s)) - \beta(X_{\lambda'}^\nu(s))|_2^2 ds$$

$$\leq 2(\lambda + \lambda')\mathbb{E}\int_0^t (|X_\lambda^\nu(s)|_2^2 + |X_{\lambda'}^\nu(s)|_2^2)ds + C_\infty^2 \int_0^t |X_\lambda^\nu(s) - X_{\lambda'}^\nu(s)|_{-1,\nu}^2 ds$$

$$+ 2\mathbb{E}\left(\int_0^t \sum_{k=1}^\infty \mu_k^2 \langle X_\lambda^\nu(s) - X_{\lambda'}^\nu(s), (X_\lambda^\nu(s) - X_{\lambda'}^\nu(s))e_k\rangle_{-1,\nu}^2 ds\right)^{1/2}.$$

By (6.17) and Young's inequality, the latter term is dominated by

$$\frac{1}{2}\mathbb{E}\left[\sup_{s\in[0,t]} \|X_\lambda^\nu(s) - X_{\lambda'}^\nu(s)\|_{-1,\nu}^2\right] + \frac{1}{2}C_\infty^2\mathbb{E}\int_0^t \|X_\lambda^\nu(s) - X_{\lambda'}^\nu(s)\|_{-1,\nu}^2 ds.$$

Hence, because of $x \in L^2(\mathbb{R}^d)$ and Claim 1, we may now apply Gronwall's lemma to obtain that, for some constant C independent of λ', λ (and ν),

$$\mathbb{E}\left[\sup_{t\in[0,T]} \|X_\lambda^\nu(t) - X_{\lambda'}^\nu(t)\|_{-1,\nu}^2\right] + \mathbb{E}\int_0^T |\beta(X_\lambda^\nu(s)) - \beta(X_{\lambda'}^\nu(s))|_2^2 ds \leq C(\lambda + \lambda').$$

$$(6.34)$$

Hence there exists an $(\mathscr{F}_t)$-adapted continuous H^{-1}-valued process $X^\nu = (X^\nu(t))_{t\in[0,T]}$ such that $X^\nu \in L^2(\Omega; C([0, T]; H^{-1}))$. Now, by Claim 1, it follows that

$$X^\nu \in L^2([0, T] \times \Omega; L^2(\mathbb{R}^d)).$$

Claim 2 X^ν satisfies Eq. (6.23) (i.e., we can pass to the limit in (6.27) as $\lambda \to 0$).

We already know that

$$X_\lambda^\nu \longrightarrow X^\nu \quad \text{and} \quad \int_0^\cdot X_\lambda^\nu(s)dW(s) \longrightarrow \int_0^\cdot X^\nu(s)dW(s)$$

in $L^2(\Omega; C([0, T]; H^{-1}))$ as $\lambda \to 0$ (for the second convergence see the above argument using (6.17) and inequality (1.23)). So, by (6.27) it follows that

$$\int_0^\cdot (\beta(X_\lambda^\nu(s)) + \lambda X_\lambda^\nu(s)))ds, \quad \lambda > 0,$$

converges as $\lambda \to 0$ to an element in $L^2(\Omega; C([0, T]; H^1))$. But, by (6.34) and Claim 1, it follows that

$$\int_0^{\bullet} (\beta(X_\lambda^\nu(s)) + \lambda X_\lambda^\nu(s))ds \longrightarrow \int_0^{\bullet} \beta(X^\nu(s))ds \qquad (6.35)$$

as $\lambda \to 0$ in $L^2(\Omega; L^2([0, T]; L^2(\mathbb{R}^d)))$. Hence Claim 2 is proved.

Now, (6.24) follows from Claim 1 by lower semicontinuity. This completes the proof of Lemma 6.3.4. $\square$

Now we can end the proof of Theorem 6.3.3. We are going to use Lemma 6.3.4 and let $\nu \to 0$. The arguments are similar to those in the proof of Lemma 6.3.4. So, we shall not repeat all the details.

Now, we rewrite (6.21) as

$$dX^\nu + (I - \Delta)\beta(X^\nu)dt = (1 - \nu)\beta(X^\nu)dt + X^\nu dW(t) \qquad (6.36)$$

and apply Itô's formula to $\varphi(x) = \frac{1}{2}\|x\|_{-1}^2$. We get, for $x \in H^{-1}$, by (6.18) and after taking expectation,

$$\frac{1}{2}\mathbb{E}\|X^\nu(t)\|_{-1}^2 + \mathbb{E}\int_0^t \int_{\mathbb{R}^d} \beta(X^\nu(s))X^\nu(s)d\xi\, ds$$

$$= \frac{1}{2}\|x\|_{-1}^2 + (1 - \nu)\mathbb{E}\int_0^t \langle\beta(X^\nu(s)), X^\nu(s)\rangle_{-1}ds$$

$$+ \frac{1}{2}\mathbb{E}\int_0^t \sum_{k=1}^\infty \mu_k^2\|X^\nu e_k\|_{-1}^2 ds$$

$$\leq \frac{1}{2}|x|_{-1}^2 + \mathbb{E}\int_0^t \|\beta(X^\nu)\|_{-1}\|X^\nu\|_{-1}ds$$

$$+ \frac{1}{2}C_\infty^2 \mathbb{E}\int_0^t \|X^\nu(s)\|_{-1}^2 ds, \quad \forall t \in [0, T].$$

Recalling that $\|\cdot\|_{-1} \leq |\cdot|_2$, we get, via Young's and Gronwall's inequalities, for some $C \in (0, \infty)$ that

$$\mathbb{E}\|X^\nu(t)\|_{-1}^2 + \frac{\alpha}{2}\mathbb{E}\int_0^T |X^\nu(s)|_2^2 ds \leq C\|x\|_{-1}^2, \ t \in [0, T], \ \nu \in (0, 1), \qquad (6.37)$$

because, by Hypothesis 8(i), $\beta(r)r \geq \overline{\alpha}|\beta(r)|^2, \forall r \in \mathbb{R}$, with $\overline{\alpha} := (\beta + 1)^{-1}$. Here we set $\alpha = 0$ if (6.20) does not hold.

Now, by a similar calculus, for $X^\nu - X^{\nu'}$ we get for $t \in [0, T]$

$$\|X^\nu(t) - X^{\nu'}(t)\|_{-1}^2 + 2 \int_0^t \int_{\mathbb{R}^d} (\beta(X^\nu) - \beta(X^{\nu'}))(X^\nu - X^{\nu'}) d\xi \, ds$$

$$\leq C \int_0^t \langle \beta(X^\nu) - \beta(X^{\nu'}), X^\nu - X^{\nu'} \rangle_{-1} ds$$

$$+ C \int_0^t (\nu |\beta(X^\nu)|_2 + \nu' |\beta(X^{\nu'})|_2) \|X^\nu - X^{\nu'}\|_{-1} ds$$

$$+ C \int_0^t \|X^\nu - X^{\nu'}\|_{-1}^2 ds + \sum_{k=1}^\infty \int_0^t \mu_k \langle (X^\nu - X^{\nu'}), e_k (X^\nu - X^{\nu'}) dW_k \rangle_{-1}.$$

Taking into account that, by Hypothesis 8(i),

$$(\beta(x) - \beta(y))(x - y) \geq \bar{\alpha} |\beta(x) - \beta(y)|^2, \ \forall \, x, y \in \mathbb{R}^d,$$

we get, for all $\nu, \nu' > 0, t \in [0, T]$,

$$\|X^\nu(t) - X^{\nu'}(t)\|_{-1}^2 + \tilde{\alpha} \int_0^t |\beta(X^\nu(s)) - \beta(X^{\nu'}(s))|_2^2 ds$$

$$\leq C_1 \int_0^t \|X^\nu(s) - X^{\nu'}(s)\|_{-1}^2 ds + \frac{\tilde{\alpha}}{2} \int_0^t |\beta(X^\nu(s)) - \beta(X^{\nu'}(s))|_2^2 ds$$

$$+ C_2(\nu + \nu') \int_0^t (|\beta(X^\nu(s))|_2^2 + |\beta(X^{\nu'}(s))|_2^2) ds$$

$$+ \sum_{k=1}^\infty \int_0^t \mu_k \langle (X^\nu(s) - X^{\nu'}(s)), e_k (X^\nu(s) - X^{\nu'}(s)) \rangle_{-1} d\beta_k(s).$$

So, as in the proof of Lemma 6.3.4, by (6.24), if $x \in L^2(\mathbb{R}^d)$, and by (6.37), if $x \in H^{-1}(\mathbb{R}^d)$ and β satisfies (6.20), by (1.23) for $p = 1$, we get, for all $\nu, \nu' \in (0, 1)$,

$$\mathbb{E} \sup_{t \in [0,T]} \|X^\nu(t) - X^{\nu'}(t)\|_{-1}^2 + \mathbb{E} \int_0^T |\beta(X^\nu(s)) - \beta(X^{\nu'}(s))|_2^2 ds \leq C(\nu + \nu').$$

The remaining part of the proof is now exactly the same as the last part of the proof of Lemma 6.3.4. As regards the uniqueness of solutions it is an immediate consequence of Itô's formula, but we omit the details. □

6.4   Equation (6.2) for Maximal Monotone Functions β with Polynomial Growth

In this section, we assume $d \geq 3$ and we shall study the existence for Eq. (6.2) under the following assumptions:

Hypothesis 9

(i) $\beta : \mathbb{R} \to 2^{\mathbb{R}}$ is a maximal monotone graph such that $0 \in \beta(0)$ and

$$\sup\{|\eta|;\ \eta \in \beta(r)\} \leq C(1 + |r|^m), \quad \forall r \in \mathbb{R}, \tag{6.38}$$

where $1 \leq m < \infty$.

(ii) $W(t) = \sum_{k=1}^{\infty} \mu_k e_k W_k(t), \quad t \geq 0$, where $\{W_k\}$ are independent Brownian motions on a filtered probability space $\{\Omega, \mathscr{F}, \mathscr{F}_t, \mathbb{P}\}$, $\mu_k \in \mathbb{R}$, and $e_k \in C^1(\mathbb{R}^d) \cap \mathscr{H}^{-1}$ are such that $\{e_k\}$ is an orthonormal basis in $\mathscr{H}^{-1}$ and

$$\sum_{k=1}^{\infty} \mu_k^2(|e_k|_\infty^2 + |\nabla e_k|_d^2 + 1) < \infty. \tag{6.39}$$

The existence of $\{e_k\}$ as in Hypothesis 9(ii) is ensured by the following lemma.

Lemma 6.4.1 Let $d \geq 3$ and let $e \in L^\infty(\mathbb{R}^d; \mathbb{R}^d)$ be such that $\nabla e \in L^d(\mathbb{R}^d; \mathbb{R}^d)$. Then

$$|xe|_{-1} \leq |x|_{-1}(|e|_\infty + C|\nabla e|_d), \quad \forall x \in \mathscr{H}^{-1}, \tag{6.40}$$

where C is independent of x and e.

Proof We have

$$|xe|_{-1} = \sup\{\langle x, e\varphi\rangle; |\varphi|_1 \leq 1\} \leq |x|_{-1} \sup\{|e\varphi|_1; |\varphi|_1 \leq 1\}. \tag{6.41}$$

On the other hand, by Lemma 6.2.1 we have, for all $\varphi \in C_0^\infty(\mathbb{R}^d)$,

$$|e\varphi|_1 \leq |e\nabla\varphi + \varphi\nabla e|_2 \leq |e\nabla\varphi|_2 + |\varphi\nabla e|_2$$

$$\leq |e|_\infty|\nabla\varphi|_2 + |\varphi|_p|\nabla e|_d \leq |e|_\infty|\varphi|_1 + C|\varphi|_1|\nabla e|_d,$$

where $p = \frac{2d}{d-2}$. Then, by (6.41), (6.40) follows, as claimed. $\square$

Remark 6.4.2

(i) It should be mentioned that, for $d = 2$, Lemma 6.4.1 fails and this is the main reason why our treatment of Eq. (6.2) under Hypothesis 9(i)(ii) is constrained to $d \geq 3$.
(ii) We note that Remark 6.3.2 with the role of $H^{-1}(\mathbb{R}^d)$ replaced by $\mathscr{H}^{-1}$ remains true in all its parts under condition Hypothesis 9(ii) above.   □

We denote by $j : \mathbb{R} \to \mathbb{R}$ the potential associated with β (See Sect. 1.2.4).

Definition 6.4.3 Let $x \in \mathscr{H}^{-1}$ and $p := \max(2, 2m)$. An $\mathscr{H}^{-1}$-valued adapted process $X = X(t)$ is called strong solution to (6.2) if the following conditions hold

(i) X is $\mathscr{H}^{-1}$-valued continuous on $[0, T]$, $\mathbb{P}$-a.s.
(ii) $X \in L^p(\Omega \times (0, T) \times \mathbb{R}^d)$.
(iii) There is $\eta \in L^{\frac{p}{m}}(\Omega \times (0, T) \times \mathbb{R}^d)$ such that $\eta \in \beta(X)$, $dt \otimes \mathbb{P} \otimes d\xi$-a.e. on $(0, T) \times \Omega \times \mathbb{R}^d$.
(iv) We have $\mathbb{P}$a.s.

$$X(t) = x + \Delta \int_0^t \eta(s)ds + \sum_{k=1}^\infty \mu_k \int_0^t X(s)e_k dW_k(s), \quad \text{in } \mathscr{D}'(\mathbb{R}^d). \quad (6.42)$$

Theorem 6.4.4 below is the main existence result for Eq. (6.2). It can be compared with Theorem 3.4.1.

Theorem 6.4.4 *Assume that $d \geq 3$ and that $x \in L^p(\mathbb{R}^d) \cap L^2(\mathbb{R}^d) \cap \mathscr{H}^{-1}$, $p := \max(2, 2m)$. Then, under Hypothesis 9(i)(ii) there is a unique strong solution X to (6.2) such that*

$$X \in L^2(\Omega; C([0, T]; \mathscr{H}^{-1})). \quad (6.43)$$

Moreover, if $x \geq 0$, a.e. in $\mathbb{R}^d$, then $X \geq 0$, a.e. on $(0, T) \times \mathbb{R}^d \times \Omega$.

Theorem 6.4.4 is applicable to a large class of nonlinearities $\beta : \mathbb{R} \to 2^{\mathbb{R}}$ and, in particular, to

$$\beta(r) = \rho H(r) + \alpha r, \quad \beta(r) = \rho H(r - r_c)r, \quad \forall r \in \mathbb{R},$$

where $\rho > 0$, $\alpha, r_c \geq 0$, which models the dynamics of self-organized criticality (see Sect. 1.1).

Proof Consider the approximating equation

$$\begin{cases} dX_\lambda - \Delta(\beta_\lambda(X_\lambda) + \lambda X_\lambda)dt = X_\lambda dW, \quad t \in (0, T), \\ X_\lambda(0) = x, \end{cases} \quad (6.44)$$

where $\beta_\lambda = \frac{1}{\lambda}(1 - (1 + \lambda\beta)^{-1})$, $\lambda > 0$. We note that $\beta_\lambda = \nabla j_\lambda$, where (see, (1.40))

$$j_\lambda(r) = \inf\left\{\frac{|r - \bar{r}|^2}{2\lambda} + j(\bar{r}); \bar{r} \in \mathbb{R}\right\}, \quad \forall\, r \in \mathbb{R}.$$

We need the following.

Lemma 6.4.5 *Let $x \in \mathscr{H}^{-1} \cap L^p(\mathbb{R}^d) \cap L^2(\mathbb{R}^d)$, $p := 2m$, $d \geq 3$. Then (6.44) has a unique solution*

$$X_\lambda \in L^2(\Omega; C([0, T]; \mathscr{H}^{-1})) \cap L^\infty([0, T]; L^p(\Omega \times \mathbb{R}^d)). \tag{6.45}$$

Moreover, for all $\lambda, \mu > 0$, we have

$$\mathbb{E}\sup_{0 \leq t \leq T} |(X_\lambda(t) - X_\mu(t))|_{-1}^2 \leq C(\lambda + \mu), \tag{6.46}$$

$$\mathbb{E}|X_\lambda(t)|_p^p \leq C|x|_p^p, \quad \forall\, t \in [0, T], \tag{6.47}$$

$$\mathbb{E}\int_0^T \int_{\mathbb{R}^d} |\beta_\lambda(X_\lambda)|^{\frac{p}{m}} dt\, d\xi \leq C|x|_p^p, \quad \forall\, \lambda > 0, \tag{6.48}$$

$$\mathbb{E}\left[\sup_{0 \leq t \leq T} |X_\lambda(t)|_{-1}^2\right] \leq C\|x\|_{-1}^2, \quad \forall\, \lambda > 0, \tag{6.49}$$

where C is independent of λ, μ.

Proof We consider for each fixed λ the equation (see (6.21))

$$\begin{cases} dX_\lambda^\nu + (\nu - \Delta)(\beta_\lambda(X_\lambda^\nu) + \lambda X_\lambda^\nu)dt = X_\lambda^\nu dW \\[2mm] X_\lambda^\nu(0) = x, \end{cases} \tag{6.50}$$

where $\nu > 0$. Let $x \in L^2(\mathbb{R}^d) \cap L^p(\mathbb{R}^d) \cap \mathscr{H}^{-1}$. By Claim 1 in the proof of Lemma 6.3.4, (6.50) has a unique solution

$$X_\lambda^\nu \in L^2(\Omega; L^\infty([0, T]; L^2(\mathbb{R}^d))) \cap L^2(\Omega \times [0, T]; H^1(\mathbb{R}^d))$$

with continuous sample paths in $L^2(\mathbb{R}^d)$.

As seen in the proof of Theorem 6.3.3, we have, for $\nu \to 0$, $X_\lambda^\nu \to X_\lambda$ strongly in $L^2(\Omega; C([0, T]; H^{-1}(\mathbb{R}^d)))$ and, by (6.24), along a subsequence also, weakly* in $L^2(\Omega; L^\infty([0, T]; L^2(\mathbb{R}^d)))$, where X_λ is the solution to (6.44). It remains to be shown that X_λ satisfies (6.45)–(6.49). In order to explain the ideas, we apply first (formally) Itô's formula to (6.50) for the function $\varphi(x) = \frac{1}{p}|x|_p^p$. We obtain

$$
\frac{1}{p}\mathbb{E}|X_\lambda^\nu(t)|_p^p + \mathbb{E}\int_0^t\int_{\mathbb{R}^d}(\nu - \Delta)(\beta_\lambda(X_\lambda^\nu) + \lambda X_\lambda^\nu)|X_\lambda^\nu|^{p-2}X_\lambda^\nu\,ds\,d\xi
$$
$$
= \frac{1}{p}|x|_p^p + \frac{p-1}{2}\mathbb{E}\int_0^t\int_{\mathbb{R}^d}\sum_{k=1}^\infty \mu_k^2|X_\lambda^\nu e_k|^2|X_\lambda^\nu|^{p-2}dt\,d\xi. \tag{6.51}
$$

Taking into account that $X_\lambda^\nu, \beta_\lambda(X_\lambda^\nu) \in L^2(0,T;H^1(\mathbb{R}^d))$, $\mathbb{P}$-a.e, by Claim 1 in the proof of Lemma 6.3.4, we have

$$
\int_0^t\int_{\mathbb{R}^d}(\nu - \Delta)(\beta_\lambda(X_\lambda^\nu) + \lambda X_\lambda^\nu)|X_\lambda^\nu|^{p-2}X_\lambda^\nu\,ds\,d\xi
$$

$$
\geq \lambda(p-1)\int_0^t\int_{\mathbb{R}^d}|\nabla X_\lambda^\nu|^2|X_\lambda^\nu|^{p-2}d\xi\,ds,
$$

and by (6.39) we have

$$
\mathbb{E}\int_0^t\int_{\mathbb{R}^d}\sum_{k=1}^\infty \mu_k^2|X_\lambda^\nu e_k|^2|X_\lambda^\nu|^{p-2}ds\,d\xi \leq C_\infty\mathbb{E}\int_0^t\int_{\mathbb{R}^d}|X_\lambda^\nu|^p d\xi\,ds < \infty.
$$

Then, we obtain by (6.51) via Gronwall's lemma

$$
\mathbb{E}|X_\lambda^\nu(t)|_p^p \leq C|x|_p^p, \ t \in (0,T), \tag{6.52}
$$

and, by (6.38),

$$
\mathbb{E}\int_0^t\int_{\mathbb{R}^d}|\beta_\lambda(X_\lambda^\nu)|^{\frac{p}{m}}dt\,d\xi \leq C|x|_p^p, \ t \in [0,T]. \tag{6.53}
$$

It should be said, however, that the above argument is formal, because the function φ is not of class C^2 on $L^2(\mathbb{R}^d)$ and we do not know a priori if the integral in the left side of (6.51) makes sense, that is, whether $|X_\lambda^\nu|^{p-2}X_\lambda^\nu \in L^2(0,T;L^2(\Omega;H^1(\mathbb{R}^d)))$. To make it rigorous, we approximate X_λ^ν by a sequence $\{X_\lambda^{\nu,\varepsilon}\}$ of solutions to the equation

$$
\begin{cases} dX_\lambda^{\nu,\varepsilon} + A_\lambda^{\nu,\varepsilon}(X_\lambda^{\nu,\varepsilon})dt = X_\lambda^{\nu,\varepsilon}dW, \\[2mm] X_\lambda^{\nu,\varepsilon}(0) = x. \end{cases} \tag{6.54}
$$

Here, $A_\lambda^{\nu,\varepsilon} = \frac{1}{\varepsilon}(I - (I + \varepsilon A_\lambda^\nu)^{-1})$, $\varepsilon \in (0,1)$, is the Yosida approximation of the operator $A_\lambda^\nu x = (\nu - \Delta)(\beta_\lambda(x) + \lambda x)$, $\forall \ x \in D(A_\lambda^\nu) = H^1(\mathbb{R}^d)$. We set $J_\varepsilon = (I + \varepsilon A_\lambda^\nu)^{-1}$ and note that J_ε is Lipschitz in $H = H^{-1}(\mathbb{R}^d)$ as well as in all $L^q(\mathbb{R}^d)$

for $1 < q < \infty$. Moreover, we have

$$|J_\varepsilon(x)|_q \le |x|_q, \quad \forall\, x \in L^q(\mathbb{R}^d). \tag{6.55}$$

(See [21, Lemma 3.1].) Since $A_\lambda^{\nu,\varepsilon}$ is Lipschitz in H, Eq. (6.13) has a unique adapted solution $X_\lambda^{\nu,\varepsilon} \in L^2(\Omega; C([0,T]; H)$ and by Itô's formula we have

$$\frac{1}{2}\mathbb{E}\|X_\lambda^{\nu,\varepsilon}(t)\|_{-1}^2 \le \frac{1}{2}\|x\|_{-1}^2 + C_1 \sum_{k=1}^{\infty} \mu_k^2 \mathbb{E}\int_0^t \|X_\lambda^{\nu,\varepsilon}(s)e_k\|_{-1}^2 ds,$$

which, by virtue of Hypothesis 9(ii), yields

$$\mathbb{E}\|X_\lambda^{\nu,\varepsilon}(t)\|_{-1}^2 \le C_2\|x\|_{-1}^2, \quad \forall\, \varepsilon > 0,\ x \in H. \tag{6.56}$$

Similarly, since $A_\lambda^{\nu,\varepsilon}$ is Lipschitz in $L^2(\mathbb{R}^d)$ (see Lemma 6.4.6 below), we have also that $X_\lambda^{\nu,\varepsilon} \in L^2(\Omega; C([0,T]; L^2(\mathbb{R}^2)))$ and, again by Itô's formula applied to the function $|X_\lambda^{\nu,\varepsilon}(t)|_2^2$, we obtain that

$$\mathbb{E}|X_\lambda^{\nu,\varepsilon}(t)|_2^2 \le \frac{1}{2}|x|_2^2 + C_3 \sum_{k=1}^{\infty} \mu_k^2 \mathbb{E}\int_0^{\infty} |X_\lambda^{\nu,\varepsilon}(s)e_k|_2^2 ds,$$

which yields, by virtue of Hypothesis 9(ii),

$$\mathbb{E}|X_\lambda^{\nu,\varepsilon}(t)|_2^2 \le C_4|x|_2^2, \quad \forall\, t \in [0,T]. \tag{6.57}$$

Claim 1 For $p \in [2,\infty)$ and $x \in L^p(\mathbb{R}^d)$, we have that $X_\lambda^{\nu,\varepsilon} \in L_W^{\infty}([0,T];$ $L^p(\Omega; L^p(\mathbb{R}^d)) \cap L^2(\Omega; L^2(\mathbb{R}^d)))$, where here and below the subscript W refers to $(\mathscr{F}_t)$-adapted processes.

For $R > 0$, consider the set

$$\mathscr{K}_R = \{X \in L_W^{\infty}([0,T]; L^p(\Omega; L^p(\mathbb{R}^d)) \cap L^2(\Omega; L^2(\mathbb{R}^d))),$$

$$e^{-p\alpha t}\mathbb{E}|X(t)|_p^p \le R^p,\ e^{-2\alpha t}\mathbb{E}|X(t)|_2^2 \le R^2, t \in [0,T]\}.$$

Since, by (6.54), $X_\lambda^{\nu,\varepsilon}$ is a fixed point of the map

$$F: X \mapsto e^{-\frac{t}{\varepsilon}}X + \frac{1}{\varepsilon}\int_0^t e^{-\frac{t-s}{\varepsilon}}J_\varepsilon(X(s))ds + \int_0^t e^{-\frac{(t-s)}{\varepsilon}}X(s)dW(s),$$

obtained by iteration in $C_W([0, T]; L^2(\Omega; L^2(\mathbb{R}^d)))$, it suffices to show that F leaves the set $\mathcal{K}_R$ invariant for $R > 0$ large enough. By (6.55), we have

$$\left(e^{-p\alpha t} \mathbb{E} \left| e^{-\frac{t}{\varepsilon}} x + \frac{1}{\varepsilon} \int_0^t e^{-\frac{t-s}{\varepsilon}} J_\varepsilon(X(s)) ds \right|_p^p \right)^{\frac{1}{p}}$$

$$\leq e^{-\left(\frac{1}{\varepsilon} + \alpha\right)t} |x|_p + e^{-\alpha t} \int_0^t \frac{1}{\varepsilon} e^{-\frac{(t-s)}{\varepsilon}} \left(\mathbb{E} |X(s)|_p^p \right)^{\frac{1}{p}} ds \qquad (6.58)$$

$$\leq e^{-\left(\frac{1}{\varepsilon} + \alpha\right)t} |x|_p + \frac{R}{1 + \alpha\varepsilon},$$

and, similarly, that

$$\left(e^{-2\alpha t} \mathbb{E} \left| e^{-\frac{t}{\varepsilon}} x + \frac{1}{\varepsilon} \int_0^t e^{-\frac{(t-s)}{\varepsilon}} J_\varepsilon(X(s)) ds \right|_2^2 \right)^{\frac{1}{2}}$$

$$(6.59)$$

$$\leq e^{-\left(\frac{1}{\varepsilon} + \alpha\right)t} |x|_2 + \frac{R}{1 + \alpha\varepsilon}.$$

Now, we set

$$Y(t) = \int_0^t e^{-\frac{(t-s)}{\varepsilon}} X(s) dW(s), \ t \geq 0.$$

We have

$$\begin{cases} dY + \dfrac{1}{\varepsilon} Y \, dt = X \, dW, & \forall \, t \geq 0, \\[2mm] Y(0) = 0. \end{cases}$$

Equivalently,

$$d(e^{\frac{t}{\varepsilon}} Y(t)) = e^{\frac{t}{\varepsilon}} X(t) dW(t), \ t > 0; \quad Y(0) = 0.$$

By Proposition 1.2.3, it follows that $e^{\frac{t}{\varepsilon}} Y$ is an $L^p(\mathbb{R}^d)$-valued $(\mathscr{F}_t)$-adapted continuous process on $[0, \infty)$ and

$$\mathbb{E} |e^{\frac{t}{\varepsilon}} Y(t)|_p^p = \frac{1}{2} p(p-1) \sum_{k=1}^{\infty} \mu_k^2 \mathbb{E} \int_0^t \int_{\mathbb{R}^d} |e^{\frac{s}{\varepsilon}} Y(s)|^{p-2} |e^{\frac{s}{\varepsilon}} X(s) e_k|^2 ds.$$

This yields via Hypothesis 9(ii)

$$\mathbb{E}|e^{\frac{t}{\varepsilon}}Y(t)|_p^p \leq \frac{1}{2}(p-1)\mathbb{E}\int_0^t |e^{\frac{s}{\varepsilon}}Y(s)|_p^p ds + C\mathbb{E}\int_0^t |e^{\frac{s}{\varepsilon}}X(s)|_p^p ds, \quad \forall\, t \in [0, T],$$

and, therefore,

$$\mathbb{E}|Y(t)|_p^p \leq C_1 e^{-(\alpha+\frac{1}{\varepsilon})pt}\mathbb{E}\int_0^t |e^{\frac{s}{\varepsilon}}X(s)|_p^p ds \leq \frac{R^p e^{-p\alpha t}\varepsilon C_1}{p(1+\varepsilon\alpha)}, \quad \forall\, t \in [0, T].$$

Similarly, we get

$$e^{-2\alpha t}\mathbb{E}|Y(t)|_2^2 \leq \frac{R^2 \varepsilon C_1}{2(1+\varepsilon\alpha)}, \quad \forall\, t \in [0, T].$$

Then, by formulae (6.58), (6.59), we infer that, for α large enough and $R > 2(|x|_p + |x|_2)$, F leaves $\mathscr{K}_R$ invariant, which proves Claim 1.

Claim 2 We have, for all $p \in [2, \infty)$ and $x \in L^p(\mathbb{R}^d)$, that there exists $C_p \in (0, \infty)$ such that

$$\text{ess sup}_{t\in[0,T]}\mathbb{E}|X_\lambda^{\nu,\varepsilon}(t)|_p^p \leq C_p \quad \text{for all } \varepsilon, \lambda, \nu \in (0,1). \tag{6.60}$$

Again invoking Proposition 1.2.3, we have by (6.54) that $X_\lambda^{\nu,\varepsilon}$ satisfies

$$\mathbb{E}|X_\lambda^{\nu,\varepsilon}(t)|_p^p = |x|_p^p - p\,\mathbb{E}\int_0^t \int_{\mathbb{R}^d} A_\lambda^{\nu,\varepsilon}(X_\lambda^{\nu,\varepsilon})X_\lambda^{\nu,\varepsilon}|X_\lambda^{\nu,\varepsilon}|^{p-2}d\xi\,ds$$

$$+ p(p-1)\sum_{k=1}^\infty \mu_k^2 \mathbb{E}\int_0^t \int_{\mathbb{R}^d} |X_\lambda^{\nu,\varepsilon}|^{p-2}|X_\lambda^{\nu,\varepsilon}e_k|^2 d\xi\,ds. \tag{6.61}$$

On the other hand, $A_\lambda^{\nu,\varepsilon}(X_\lambda^{\nu,\varepsilon}) = \frac{1}{\varepsilon}(X_\lambda^{\nu,\varepsilon} - J_\varepsilon(X_\lambda^{\nu,\varepsilon}))$ and so we have

$$\int_{\mathbb{R}^d} A_\lambda^{\nu,\varepsilon}(X_\lambda^{\nu,\varepsilon})X_\lambda^{\nu,\varepsilon}|X_\lambda^{\nu,\varepsilon}|^{p-2}d\xi = \frac{1}{\varepsilon}\int_{\mathbb{R}^d} |X_\lambda^{\nu,\varepsilon}|^p d\xi - \frac{1}{\varepsilon}\int_{\mathbb{R}^r} J_\varepsilon(X_\lambda^{\nu,\varepsilon})|X_\lambda^{\nu,\varepsilon}|^{p-2}X_\lambda^{\nu,\varepsilon}d\xi.$$

Recalling (6.55), we get, via the Hölder inequality,

$$\int_{\mathbb{R}^d} A_\lambda^{\nu,\varepsilon}(X_\lambda^{\nu,\varepsilon})X_\lambda^{\nu,\varepsilon}|X_\lambda^{\nu,\varepsilon}|^{p-2}d\xi \geq 0,$$

and so, by (6.61) and Hypothesis 9(ii), we obtain, via Gronwall's lemma, estimate (6.60), as claimed.

Claim 3 We have, for $\varepsilon \to 0$,

$$X_\lambda^{\nu,\varepsilon} \longrightarrow X_\lambda^\nu \text{ strongly in } L_W^\infty([0,T]; L^2(\Omega; H))$$

and weakly* in $L^\infty([0,T]; L^p(\Omega; L^p(\mathbb{R}^d)) \cap L^2(\Omega; L^2(\mathbb{R}^d)))$.

For simplicity, we write X_ε instead of $X_\lambda^{\nu,\varepsilon}$ and X instead of X_λ^ν. Also, we set $\gamma(r) \equiv \beta_\lambda(r) + \lambda r$.

Subtracting Eqs. (6.54) and (6.50), we get via Itô's formula and because $A_\lambda^{\nu,\varepsilon}$ is monotone on H

$$\frac{1}{2}\,\mathbb{E}\|X_\varepsilon(t) - X(t)\|_{-1,\nu}^2 + \mathbb{E}\int_0^t \int_{\mathbb{R}^d} (\gamma(J_\varepsilon(X)) - \gamma(X))(X_\varepsilon - X)d\xi\,ds$$

$$\leq C\,\mathbb{E}\int_0^t \|X_\varepsilon(s) - X(s)\|_{-1,\nu}^2 ds,$$

and hence, by Gronwall's lemma, we obtain

$$\mathbb{E}\|X_\varepsilon(t) - X(t)\|_{-1,\nu}^2 \leq C\mathbb{E}\int_0^T \int_{\mathbb{R}^d} |\gamma(J_\varepsilon(X)) - \gamma(X)||X_\varepsilon - X|d\xi\,ds. \qquad (6.62)$$

On the other hand, it follows by (6.55) that

$$\int_{\Omega \times [0,T] \times \mathbb{R}^d} |J_\varepsilon(X)|^2 \mathbb{P}(d\omega)dt\,d\xi \leq \int_{\Omega \times [0,T] \times \mathbb{R}^d} |X|^2 \mathbb{P}(d\omega)dt\,d\xi,$$

while, for $\varepsilon \to 0$,

$$J_\varepsilon(y) \longrightarrow y \text{ in } H^{-1}, \quad \forall\, y \in H^{-1},$$

(because $A_\lambda^{\nu,\varepsilon}$ is maximal monotone in $H^{-1}(\mathbb{R}^d)$) and so, $J_\varepsilon(X(t,\omega)) \longrightarrow X(t,\omega)$ in $H^{-1}(\mathbb{R}^d)$ for all $(t,\omega) \in (0,T) \times \Omega$. Hence, as $\varepsilon \to 0$,

$$J_\varepsilon(X) \longrightarrow X \text{ weakly in } L^2(\Omega \times [0,T] \times \mathbb{R}^d), \qquad (6.63)$$

and, according to the inequality above, this implies that, for $\varepsilon \to 0$,

$$|J_\varepsilon(X)|_{L^2((0,T) \times \Omega \times \mathbb{R}^d)} \longrightarrow |X|_{L^2((0,T) \times \Omega \times \mathbb{R}^d)}.$$

Hence, $J_\varepsilon(X) \longrightarrow X$ strongly in $L^2(\Omega \times [0,T] \times \mathbb{R}^d)$ as $\varepsilon \to 0$. Now, taking into account that γ is Lipschitz, we conclude by (6.62), (6.63) and by estimates (6.57), (6.60) that Claim 3 is true.

Now, we can complete the proof of Lemma 6.4.5. Namely, letting first $\varepsilon \to 0$ and then $\nu \to \infty$ in (6.60), we get (6.47) and hence (6.48) as desired.

Now, let us prove (6.46) and (6.49). Arguing as in the proof of Theorem 6.3.3, we obtain

$$\frac{1}{2}\|X_\lambda^\nu(t)\|_{-1,\nu}^2 + \int_0^t \int_{\mathbb{R}^d} (\beta_\lambda(X_\lambda^\nu) + \lambda X_\lambda^\nu)X_\lambda^\nu d\xi \, ds$$

$$= \frac{1}{2}\|x\|_{-1,\nu}^2 + \frac{1}{2}\int_0^t \sum_{k=1}^\infty \mu_k^2\|X_\lambda^\nu e_k\|_{-1,\nu}^2 \, ds \tag{6.64}$$

$$+ \int_0^t \langle X_\lambda^\nu, X_\lambda^\nu dW \rangle_{-1,\nu}.$$

Keeping in mind that, by (6.40), $\|X_\lambda^\nu e_k\|_{-1,\nu} \leq C\|X_\lambda^\nu\|_{-1,\nu}(|e_k|_\infty + |\nabla e_k|_d)$, where C is independent of ν, we obtain by (1.23) for $p = 1$ (cf. the proof of Theorem 6.3.3)

$$\mathbb{E} \sup_{t \in [0,T]} \|X_\lambda^\nu(t)\|_{-1,\nu}^2 + \lambda \mathbb{E} \int_0^T |X_\lambda^\nu|_2^2 ds \leq C\|x\|_{-1,\nu}^2.$$

Taking into account that

$$\lim_{\nu \to 0} \|y\|_{-1,\nu} = |y|_{-1}, \quad \forall y \in \mathcal{H}^{-1},$$

we obtain, as in Theorem 6.3.3 (see the part following (6.36)), that

$$\mathbb{E} \left[\sup_{t \in [0,T]} |X_\lambda(t)|_{-1}^2 \right] + \lambda \mathbb{E} \int_0^T |X_\lambda(t)|_2^2 dt \leq C|x|_{-1}^2, \quad \forall \, \lambda > 0, \tag{6.65}$$

where C is independent of λ. In particular, (6.49) holds.

Completely similarly, one proves (6.46). Namely, we have

$$d(X_\lambda^\nu - X_\mu^\nu) + (\nu - \Delta)(\beta_\lambda(X_\lambda^\nu) + \lambda X_\lambda^\nu - \beta_\mu(X_\mu^\nu) - \mu X_\mu^\nu)dt$$

$$= (X_\lambda^\nu - X_\mu^\nu)dW$$

and again proceeding as in the proof of Theorem 6.3.3, we obtain as above that

$$\frac{1}{2}\|X_\lambda^\nu(t) - X_\mu^\nu(t)\|_{-1,\nu}^2 + \int_0^t \int_{\mathbb{R}^d} (\beta_\lambda(X_\lambda^\nu) + \lambda X_\lambda^\nu - \beta_\mu(X_\mu^\nu) - \mu X_\mu^\nu)(X_\lambda^\nu - X_\mu^\nu)d\xi \, ds$$

$$= \frac{1}{2}\int_0^t \int_{\mathbb{R}^d} \sum_{k=1}^\infty \mu_k^2 |(X_\lambda^\nu - X_\mu^\nu)e_k|_{-1,\nu}^2 ds$$

$$+ \int_0^t \langle X_\lambda^\nu - X_\mu^\nu, (X_\lambda^\nu - X_\mu^\nu)dW \rangle_{-1,\nu}, \quad t \in [0, T].$$

Then, applying once again (1.23) for $p = 1$, and the fact that, by Hypothesis 9(i), $|\beta_\lambda(r)| \le C|r|^m$, $\forall$ $r \in \mathbb{R}$ with C independent of λ, we get, proceeding as in the proof of Theorem 6.3.3, that

$$\mathbb{E}\left[\sup_{t\in[0,T]} |X^\nu_\lambda(t) - X^\nu_\mu(t)|^2_{-1} \right] \le C(\lambda + \mu),$$

where C is independent of ν, λ, μ. Letting $\nu \to 0$ as in the previous case, we obtain (6.46), as claimed. This completes the proof of Lemma 6.4.5. $\square$

Above we have used the following lemma.

Lemma 6.4.6 $A^{\nu,\varepsilon}_\lambda$ *is Lipschitz in* $L^2(\mathbb{R}^d)$.

Proof It suffices to check that J_ε is Lipschitz in $L^2(\mathbb{R}^d)$. We set $\gamma(r) = \beta_\lambda(r) + \lambda r$. We have, for $x, \bar{x} \in L^2(\mathbb{R}^d)$,

$$J_\varepsilon(x) - J_\varepsilon(\bar{x}) - \varepsilon\Delta(\gamma(J_\varepsilon(x)) - \gamma(J_\varepsilon(\bar{x}))) = x - \bar{x}.$$

Multiplying by $\gamma(J_\varepsilon(x)) - \gamma(J_\varepsilon(\bar{x}))$ in $L^2(R^d)$, we get

$$\langle J_\varepsilon(x) - J_\varepsilon(\bar{x}), \gamma(J_\varepsilon(x)) - \gamma(J_\varepsilon(\bar{x}))\rangle_2 \le |\gamma(J_\varepsilon(x)) - \gamma(J_\varepsilon(\bar{x}))|_2 |x - \bar{x}|_2.$$

Taking into account that $(\gamma(r) - \gamma(\bar{r}))(r - \bar{r}) \ge L|r - \bar{r}|$, $\forall$ $r, \bar{r} \in \mathbb{R}$, and that γ is Lipschitz, we get

$$|J_\varepsilon(x) - J_\varepsilon(\bar{x})|_2 \le C|x - \bar{x}|_2,$$

as claimed. $\square$

Now we can end the proof of Theorem 6.4.4. By (6.46)–(6.49), it follows that there is a process $X \in L^\infty([0,T]; L^p(\Omega \times \mathbb{R}^d))$ such that, for $\lambda \to 0$,

$$X_\lambda \to X \text{ weak-star in } L^\infty([0,T]; L^p(\Omega \times \mathbb{R}^d))$$

$$\lambda X_\lambda \to 0 \text{ strongly in } L^2([0,T]; L^2(\Omega \times \mathbb{R}^d))$$

$$\beta_\lambda(X_\lambda) \to \eta \text{ weakly in } L^{\frac{p}{m}}([0,T] \times \Omega \times \mathbb{R}^d) \tag{6.66}$$

$$X_\lambda \to X \text{ strongly in } L^2(\Omega; C([0,T]; \mathscr{H}^{-1})).$$

It remains to be shown that X is a strong solution to (6.2) in the sense of Definition 6.4.3.

By (6.44) and (6.66), we see that

$$\begin{cases} dX - \Delta\eta dt = XdW, \ t \in (0, T) \\ X(0) = x. \end{cases} \tag{6.67}$$

We need to show that $\eta \in \beta(X)$ a.e. but unfortunately Proposition 1.2.8 is not directly applicable in this case because $\mathcal{O} = \Omega \times [0, T] \times \mathbb{R}^d$ has not finite measure. To prove that $\eta \in \beta(X)$, a.e. in $\Omega \times (0, T) \times \mathbb{R}^d$, it suffices to show that, for each $\varphi \in C_0^\infty(\mathbb{R}^d)$, we have

$$\limsup_{\lambda \to 0} \mathbb{E} \int_0^T \int_{\mathbb{R}^d} \varphi^2 \beta_\lambda(X_\lambda) X_\lambda dt\, d\xi \le \mathbb{E} \int_0^T \int_{\mathbb{R}^d} \varphi^2 \eta X\, d\xi\, dt. \tag{6.68}$$

Indeed, we have by convexity of j_λ, that

$$\mathbb{E} \int_0^T \int_{\mathbb{R}^d} \varphi^2 \beta_\lambda(X_\lambda)(X_\lambda - Z) d\xi\, dt \ge \mathbb{E} \int_0^T \int_{\mathbb{R}^d} \varphi^2(j_\lambda(X_\lambda) - j_\lambda(Z)) d\xi\, dt,$$

for all $Z \in L^p((0, T) \times \Omega \times \mathbb{R}^d)$ and so, by (6.66) and (6.68), we see that for all $Z \in L^p((0, T) \times \Omega \times \mathbb{R}^d)$,

$$\mathbb{E} \int_0^T \int_{\mathbb{R}^d} \varphi^2(\eta(X - Z)) dt d\xi \ge \mathbb{E} \int_0^T \int_{\mathbb{R}^d} \varphi^2(j(X) - j(Z)) d\xi\, dt,$$

because, for $\lambda \to 0, j_\lambda(Z) \to j(Z)$, and $j_\lambda(X_\lambda) \to j(X)$, a.e. and so, by Fatou's lemma

$$\liminf_{\lambda \to 0} \mathbb{E} \int_0^T \int_{\mathbb{R}^d} \varphi^2 j_\lambda(X_\lambda) d\xi\, dt \ge \mathbb{E} \int_0^T \int_{\mathbb{R}^d} \varphi^2 j(X) d\xi\, dt.$$

Now, we take $\varphi \in C_0^\infty(\mathbb{R}^d)$ to be non-negative, such that $\varphi = 1$ on B_N and $\varphi = 0$, outside B_{N+1} where for a given $N \in \mathbb{N}$, B_N is the closed ball of $\mathbb{R}^d$ with radius N. We get for all $Z \in L^p((0, T) \times \Omega \times \mathbb{R}^d)$

$$\mathbb{E} \int_0^T \int_{B_{N+1}} \varphi^2(\eta(X - Z)) d\xi\, dt \ge \mathbb{E} \int_0^T \int_{\mathbb{R}^d} \varphi^2(j(X) - j(Z)) d\xi\, dt, . \tag{6.69}$$

This yields

$$\mathbb{E} \int_0^T \int_{B_{N+1}} \varphi^2 \eta(X - Z) d\xi\, dt \ge \mathbb{E} \int_0^T \int_{B_{N+1}} \varphi^2 \zeta(X - Z) d\xi\, dt, \tag{6.70}$$

for all $Z \in L^p((0,T) \times \Omega \times B_{N+1})$ and $\zeta \in L^{p'}((0,T) \times \Omega \times B_{N+1})$ such that $\zeta \in \beta(Z)$, a.e. in $(0,T) \times \Omega \times B_{N+1}$.

We denote by $\widetilde{\beta} : L^p((0,T) \times \Omega \times B_{N+1}) \to L^{p'}((0,T) \times \Omega \times B_{N+1})$ the realization of the mapping β in $L^p((0,T) \times \Omega \times B_{N+1})$, that is,

$$\widetilde{\beta}(Z) = \{\zeta \in L^{p'}((0,T) \times \Omega \times B_{N+1}), \quad \zeta \in \beta(Z), \text{ a.e.}\}.$$

Since $\frac{m}{p} \leq p'$ with $\frac{1}{p'} = 1 - \frac{1}{p}$, by virtue of Hypothesis 9(i), $\widetilde{\beta}$ is maximal monotone in $L^p((0,T) \times \Omega \times B_{N+1}) \times L^{p'}((0,T) \times \Omega \times B_{N+1})$, and so, the equation

$$J(Z) + \widetilde{\beta}(Z) \ni J(X) + \eta, \tag{6.71}$$

where $J(Z) = |Z|^{p-2}Z$, has a unique solution (Z, η) (see, e.g., [6, p. 31]).

If, in (6.70), we take Z the solution to (6.71), we obtain that

$$\mathbb{E} \int_0^T \int_{B_{N+1}} \varphi^2 (J(X) - J(Z))(X - Z) dt d\xi \leq 0.$$

Then, choosing $\alpha = \frac{2}{p}$, yields

$$\mathbb{E} \int_0^T \int_{B_{N+1}} \left(|\varphi^\alpha X|^{p-2} \varphi^\alpha X - |\varphi^\alpha Z|^{p-2} \varphi^\alpha Z \right) (\varphi^\alpha X - \varphi^\alpha Z) dt d\xi \leq 0.$$

Consequently, this gives

$$\mathbb{E} \int_0^T \int_{\mathbb{R}^d} (J(\varphi^\alpha X) - J(\varphi^\alpha Z))(\varphi^\alpha X - \varphi^\alpha Z) dt d\xi \leq 0. \tag{6.72}$$

On the other hand, we have

$$J(\varphi^\alpha X) - J(\varphi^\alpha Z) = (p-1)|\lambda \varphi^\alpha X + (1-\lambda)\varphi^\alpha Z|^{p-2}(X - Z),$$

for some $\lambda = \lambda(X, Z) \in [0,1]$. Substituting into (6.72) yields

$$|\varphi^\alpha (X - Z)|^2 = 0 \text{ a.e. in } (0,T) \times \Omega \times B_{N+1},$$

Hence, $X = Z$ on $(0,T) \times \Omega \times B_N$.

Coming back to (6.71), this gives $\eta \in \beta(X)$, $dt d\mathbb{P} d\xi$, a.e., because N is arbitrary.

(1)

$$dY_\lambda - \Delta(\varphi(\beta_\lambda(X_\lambda) + \lambda X_\lambda))dt$$
$$+2\nabla\varphi \cdot \nabla(\beta_\lambda(X_\lambda) + \lambda X_\lambda)dt + (\beta_\lambda(X_\lambda) + \lambda X_\lambda)\Delta\varphi dt$$
$$= Y_\lambda dW.$$

(2)

$$dY_\lambda^\nu + (\nu - \Delta)(\varphi(\beta_\lambda(X_\lambda^\nu) + \lambda X_\lambda^\nu))dt$$
$$+2\nabla\varphi \cdot \nabla(\beta_\lambda(X_\lambda) + \lambda X_\lambda^\nu)dt + (\beta_\lambda(X_\lambda^\nu) + \lambda X_\lambda^\nu)\Delta\varphi dt$$
$$= Y_\lambda^\nu dW.$$

(3) $dY - \Delta(\varphi\eta) + 2\nabla\varphi \cdot \nabla\eta dt + \eta\Delta\varphi dt = YdW.$

By formulae (1)–(3) we have

(4)

$$\frac{1}{2}\mathbb{E}\|Y_\lambda^\nu(t)\|_{-1}^2 + \mathbb{E}\int_0^t (-\Delta)^{-1}[(\nu - \Delta)(\varphi(\beta_\lambda(X_\lambda^\nu) + \lambda X_\lambda^X \nu)]ds$$

$$+2\mathbb{E}\int_0^t \langle(-\Delta)^{-1}[\nabla\varphi \cdot \nabla(\beta_\lambda(X_\lambda^\nu) + \lambda X_\lambda^\nu)), Y_\lambda^\nu\rangle_2\, ds$$

$$+\mathbb{E}\int_0^t \langle(-\Delta)^{-1}(\Delta\varphi(\beta_\lambda(X_\lambda^\nu) + \lambda X_\lambda^\nu)), Y_\lambda^\nu\rangle_2\, ds$$

$$+\frac{1}{2}\|\varphi x\|_{-1}^2 + \frac{1}{2}\mathbb{E}\sum_{k=1}^\infty \mu_k^2 \int_0^t \|Y_\lambda^\nu(s)e_k\|_{-1}^2\, ds$$

(5)

$$\frac{1}{2}\|Y(t)\|_{-1}^2 + \int_0^t \langle\varphi\eta, Y\rangle_2\, ds$$

$$+2\int_0^t \langle(-\Delta)^{-1}(\nabla\varphi \cdot \nabla\eta, Y\rangle_2\, ds + \int_0^t \langle(-\Delta)^{-1}(\eta\Delta\varphi), Y\rangle_2\, ds$$

$$+\frac{1}{2}\|\varphi x\|_{-1}^2 + \frac{1}{2}\mathbb{E}\sum_{k=1}^\infty \mu_k^2 \int_0^t \|Y(s)e_k\|_{-1}^2\, ds$$

$$\frac{1}{2}\mathbb{E}\|\varphi X_\lambda^\nu(t)\|_{-1}^2 + \mathbb{E}\int_0^t \langle (-\Delta)^{-1}(\nu - \Delta)(\varphi(\beta_\lambda(X_\lambda^\nu) + \lambda X_\lambda^\nu), \varphi X_\lambda^\nu)\rangle ds$$

$$\leq \frac{1}{2}\|\varphi x\|_{-1}^2 + \frac{1}{2}\mathbb{E}\int_0^t \sum_{k=1}^\infty \mu_k^2 \|\varphi X_\lambda^\nu e_k\|_{-1}^2 ds.$$

$$+\mathbb{E}\int_0^t \langle (-\Delta)^{-1}(2(\nabla\varphi \cdot \nabla(\psi_\lambda(X_\lambda^\nu) + \lambda X_\lambda^\nu), \varphi X_\lambda^\nu)_2$$

$$+\langle (-\Delta)^{-1}(\Delta\varphi(\psi_\lambda(X_\lambda^\nu) + \lambda X_\lambda^\nu), \varphi X_\lambda^\nu)_2\rangle ds$$

Then, letting $\nu \to 0$, we obtain

$$\frac{1}{2}\mathbb{E}\|\varphi X_\lambda(t)\|_{-1}^2 + \mathbb{E}\int_0^t \langle \beta_\lambda(X_\lambda) + \lambda X_\lambda, \varphi^2 X_\lambda\rangle_2 ds+$$

$$\leq \frac{1}{2}\|\varphi x\|_{-1}^2 + \frac{1}{2}\mathbb{E}\int_0^t \sum_{k=1}^\infty \mu_k^2 \|\varphi X_\lambda e_k\|_{-1}^2 ds. \tag{6.73}$$

$$+\mathbb{E}\int_0^t \langle 2(-\Delta)^{-1}(\nabla\varphi \cdot \nabla)(\psi_\lambda(X_\lambda) + \lambda X_\lambda), \varphi X_\lambda\rangle_2$$

$$+(-\Delta)^{-1}(\Delta\varphi(\psi_\lambda(X_\lambda) + \lambda X_\lambda), \varphi X_\lambda)_2) ds$$

On the other hand, by (6.67) we get similarly

$$\frac{1}{2}\mathbb{E}|\varphi X(t)|_{-1}^2 + \mathbb{E}\int_0^t \langle \eta(s), \varphi^2 X\rangle_2 ds + 2\mathbb{E}\int_0^t \langle (-\Delta)^{-1}\nabla_\varphi \cdot \nabla_\eta, \varphi X\rangle_2 ds$$

$$+\mathbb{E}\int_0^t \langle (-\Delta)^{-1}(\eta\Delta\varphi), X\varphi\rangle_2 ds$$

$$= \frac{1}{2}|\varphi x|_{-1}^2 + \frac{1}{2}\mathbb{E}\int_0^1 \sum_{k=1}^\infty \mu_k^2 |\varphi X e_k|_{-1}^2, \ t \in [0,T].$$

Comparing with (5) we obtain (6.68), as claimed.

If $x \geq 0$, a.e. in $\mathbb{R}^d$, it follows that $X \geq 0$, a.e. in $\Omega \times (0,T) \times \mathbb{R}^d$. To prove this, one applies Itô's formula in (6.44) to the function $x \mapsto |x^-|_2^2$ and get $(X_\lambda^\nu)^- = 0$, a.e. in $\Omega \times (0,T) \times \mathbb{R}^d$. Then, for $\nu \to 0$, we obtain the desired result. This completes the existence proof for $x \in L^2(\mathbb{R}^d) \cap L^p(\mathbb{R}^d) \cap \mathcal{H}^{-1}$.

Uniqueness If X_1, X_2 are two solutions, we have

$$\begin{cases} d(X_1 - X_2) - \Delta(\eta_1 - \eta_2)dt = (X_1 - X_2)dW, \ t \in (0, T), \\ (X_1 - X_2)(0) = 0, \end{cases}$$

where $\eta_i \in \beta(X_i)$, $i = 1, 2$, a.e. in $\Omega \times (0, T) \times \mathbb{R}^d$.

Applying again, as above (that is, via the approximating device) Itô's formula in $\mathscr{H}^{-1}$ to $\frac{1}{2}|\varphi(X_1 - X_2)|^2_{-1}$, where $\varphi \in C_0^\infty(\mathbb{R}^d)$, we get that

$$\frac{1}{2}d|\varphi(X_1 - X_2)|^2_{-1} - \langle \Delta(\eta_1 - \eta_2), \varphi(X_1 - X_2)\rangle_{-1}$$

$$= \frac{1}{2}\sum_{k=1}^\infty \mu_k^2 |\varphi(X_1 - X_2)e_k|^2_{-1}dt + \langle(X_1 - X_2), \varphi(X_1 - X_2)dW\rangle_{-1} = 0.$$

Note that, since $\eta_1 - \eta_2 \in L^{\frac{p}{m}}(\Omega \times (0, T) \times \mathbb{R}^d)$, we have

$$-\mathbb{E}\int_0^T \langle\Delta(\eta_1 - \eta_2), \varphi(X_1 - X_2)\rangle_{-1}dt = \mathbb{E}\int_0^T \int_{\mathbb{R}^d} (\eta_1 - \eta_2), \varphi(X_1 - X_2)dt\, d\xi \geq 0,$$

and, therefore,

$$\mathbb{E}|\varphi(X_1(t) - X_2(t))|^2_{-1} \leq C\int_0^t \mathbb{E}|\varphi(X_1 - X_2)|^2_{-1}ds, \quad \forall\, t \in [0, T],$$

and, since φ was arbitrary in $C_0^\infty(\mathbb{R}^d)$, we get $X_1 \equiv X_2$, as claimed. $\qquad\square$

Remark 6.4.7 The self-organized criticality model (6.4), that is, $\beta(r) \equiv H(r) =$ Heaviside function, which is not covered by Theorem 6.4.4 for $1 \leq d \leq 2$, can, however, be treated in the special case

$$W(t) = \sum_{j=1}^N \mu_j W_j(t), \ \mu_j \in \mathbb{R},$$

(i.e., spatially independent noise) via the rescaling transformation $X = e^W Y$, which reduces it to the random parabolic equation

$$\frac{\partial}{\partial t}Y - e^{-W}\Delta\beta(Y) + \frac{1}{2}\sum_{j=1}^N \mu_j^2 Y = 0.$$

By approximating W by a smooth $W_\varepsilon \in C^1([0, T]; \mathbb{R})$ and letting $\varepsilon \to 0$, after some calculation one concludes that the latter equation has a unique strong solution Y. We omit the details, but refer to [15, 24] for a related treatment on bounded domains. (See also Sect. 3.6)

6.5   The Finite Time Extinction for Fast Diffusions

Assume here that β satisfies Hypothesis 9(i) and that W is of the form from Hypothesis 9(ii). Moreover, assume that

$$\beta(r)r \geq \rho|r|^{m+1}, \quad \forall r \in \mathbb{R}, \tag{6.74}$$

where m is as in Hypothesis 9(i).

Theorem 6.5.1 below can be compared most closely with Theorem 3.7.3

Theorem 6.5.1 *Let $d \geq 3$ and $m = \frac{d-2}{d+2}$. Let $x \in L^{m+1}(\mathbb{R}^d) \cap L^2(\mathbb{R}^d) \cap \mathcal{H}^{-1}$ and let $X = X(t, \cdot)$ be the solution to (6.2) given by Theorem 6.4.4. We set*

$$\tau = \inf\{t \geq 0; \ |X(t, \cdot)|_{-1} = 0\}. \tag{6.75}$$

Then,

$$X(t, \cdot) = 0, \quad \forall \, t \geq \tau, \tag{6.76}$$

and

$$\mathbb{P}[\tau \leq t] \geq 1 - |x|_{-1}^{1-m} \frac{C^*}{\rho\gamma^{m+1}(1 - e^{-C^*(1-m)t})}. \tag{6.77}$$

where $\gamma^{-1} = \sup\{|u|_{-1}|u|_{m+1}^{-1}\}$ and $C^* > 0$ is independent of the initial condition x.

Proof We follow the arguments of Sect. 3.8. The basic inequality is (see (3.59))

$$|X(t)|_{-1}^{1-m} + \rho(1-m)\gamma^{m+1} \int_r^t \mathbb{1}_{[|X(s)|_{-1} > 0]} ds$$

$$\leq \|X(r)\|_{-1}^{1-m} + C^*(1-m) \int_r^t |X(s)|_{-1}^{1-m} ds$$

$$+ (1-m) \int_r^t \langle |X(s)|_{-1}^{(m+1)} X(s), X(s) dW(s)\rangle_{-1}, \ \mathbb{P}\text{-a.s. }, \ 0 < r < t < \infty, \tag{6.78}$$

where C^* is a suitable constant. (We note that, by virtue of (6.12), $\gamma^{-1} < \infty$.) To get (6.78), we apply the Itô formula in (6.44) to the semimartingale $|X_\lambda(t)|^2_{-1}$ and to the function $\varphi_\varepsilon(r) = (r + \varepsilon^2)^{\frac{1-m}{2}}, r > -\varepsilon^2$. We have

$$d\varphi_\varepsilon(|X_\lambda(t)|^2_{-1}) + (1 - m)(|X_\lambda(t)|^2_{-1} + \varepsilon^2)^{-\frac{m+1}{2}} \langle X_\lambda(t), \beta_\lambda(X_\lambda(t)) + \lambda X_\lambda(t)\rangle_2 dt$$

$$= \frac{1}{2}\sum_{k=1}^{\infty} \mu_k^2 \left[\frac{(1-m)|X_\lambda(t)e_k|^2_{-1}}{(|X_\lambda(t)|^2_{-1}+\varphi^2)^{\frac{m+1}{2}}} - (1-m^2)\frac{|X_\lambda(t)e_k|^2_{-1}|X_\lambda(t)|^2_{-1}}{(|X_\lambda(t)|^2_{-1}+\varphi^2)^{\frac{m+1}{2}}}\right]dt$$

$$+2\langle\varphi'_\varepsilon(|X_\lambda(t)|^2_{-1})X_\lambda(t), X_\lambda(t)dW(t)\rangle.$$

This yields

$$\varphi_\varepsilon(|X_\lambda(t)|^2_{-1}) + \rho(1 - m)\int_r^t (|X_\lambda(s)|^2_{-1} + \varepsilon^2)^{-\frac{m+1}{2}}\int_{\mathbb{R}^d}|X_\lambda|^{m+1}ds\,d\xi$$

$$\leq \varphi_\varepsilon(\|X_\lambda(r)\|^2_{-1}) + C^*\int_r^t |\varphi^{\frac{1}{2}}X_\lambda(s)|^2_{-1}(|X_\lambda(s)|^2_{-1} + \varepsilon^2)^{-\frac{1+m}{2}}ds$$

$$+2\int_r^t \langle\varphi_\varepsilon(|X_\lambda(s)|^2_{-1})X_\lambda(s), X_\lambda(s)dW(s)\rangle_{-1}.$$

Now, letting $\lambda \to 0$, we obtain that X satisfies the estimate

$$\varphi_\varepsilon(|X(t)|^2_{-1}) + \rho(1 - m)\int_r^t \left(|X(s)|^2_{-1} + \varepsilon^2\right)^{-\frac{m+1}{2}}\int_{\mathbb{R}^d}|X(s,\xi)|^{m+1}d\xi\right)ds$$

$$\leq \varphi_\varepsilon(|\widetilde{X}(t)|^2_{-1}) + C^*\int_r^t |X(s)|^2_{-1}(|\widetilde{X}(s)|^2_{-1} + \varepsilon^2)^{\frac{m+1}{2}}ds \qquad (6.79)$$

$$+2\int_r^t \langle\varphi'_\varepsilon(|X(s)|^2_{-1})X(s), X(s)dW(s)\rangle_{-1}.$$

Here, we have used the fact that, by Lemma 6.4.5, for $\lambda \to 0$,

$$X_\lambda \to X \text{ in } \mathscr{H}^{-1},$$

and, by (6.66) it follows, via Fatou's lemma,

$$\liminf_{\lambda\to 0}\int_{\mathbb{R}^d}|X_\lambda|^{m+1}d\xi \geq \int_{\mathbb{R}^d}|X|^{m+1}d\xi,$$

and

$$(|X(t)|^2_{-1} + \varepsilon^2)^{\frac{1-m}{2}} + \rho(1-m)\gamma^{m+1} \int_r^t (|X(s)|^2_{-1} + \varepsilon^2)^{-\frac{m+1}{2}} |X(s)|^{m+1}_{-1} ds$$

$$\leq (|X(r)|^2_{-1} + \varepsilon^2)^{\frac{1-m}{2}} + C^* \int_r^t |X(s)|^2_{-1}(|X(s)|^2_{-1} + \varepsilon^2)^{-\frac{m+1}{2}} ds$$

$$+2 \int_r^t \langle \varphi'_\varepsilon(|X(s)|^2_{-1}X(s), X(s)dW(s)\rangle_{-1}, \; 0 \leq r \leq t < \infty,$$

because, by (6.11), $|x|_{-1} \leq \gamma^{-1}|x|_{m+1}$ for all $x \in L^{m+1}(\mathbb{R}^d)$. Letting $\varepsilon \to 0$, we get (6.78), as claimed.

Now, we conclude the proof as in Theorem 3.7.3. Namely, by (6.78), it follows that

$$e^{-C^*(1-m)t}|X(t)|^{1-m}_{-1} + \rho(1-m)\gamma^{m+1} \int_r^t e^{-C^*(1-m)s} \mathbb{1}_{[|X_s|_{-1}>0]} ds$$

$$\leq e^{-C^*(1-m)r}|X(r)|^{1-m}_{-1}$$

$$+(1-m) \int_r^t e^{C^*(1-m)s} \langle |X(s)|^{-(m+1)}_{-1} X(s), X(s)dW(s)\rangle_{-1}$$

and, therefore, $t \to e^{-C^*(1-m)t}|X(t)|^{1-m}_{-1}$ is an $\{\mathscr{F}_t\}_{t\geq 0}$ supermartingale. Hence, $|X(t)|_{-1} = 0$ for $t \geq \tau$, because of [85, Proposition 3.4, Chap. 2]. Moreover, taking expectation for $r = 0$, we get

$$e^{-C^*(1-m)t}\mathbb{E}|X(t)|^{1-m}_{-1} + \rho(1-m)\gamma^{m+1} \int_0^t e^{-C^*(1-m)s}\mathbb{P}(\tau > s)ds \leq |x|^{1-m}_{-1}.$$

This implies that

$$\mathbb{P}(\tau > t)\frac{1 - e^{-C^*(1-m)t}}{C^*(1-m)} \leq \int_0^t e^{-C^*(1-m)s}\mathbb{P}(\tau > s)ds \leq \frac{|x|^{1-m}_{-1}}{\rho(1-m)\gamma^{m+1}},$$

and so (6.77) follows. This completes the proof.   □

Corollary 6.5.2 *Let* $x \in \mathcal{H}^{-1} \cap L^{m+1}(\mathbb{R}^d) \cap L^2(\mathbb{R}^d)$ *be such that* $|x|_{-1} < \frac{\rho\gamma^{m+1}}{C^*}$. *Let* τ *be the stopping time defined in (6.75). Then* $\mathbb{P}(\tau < \infty) > 0$. *In other words, there is extinction in finite time with positive probability.*

6.6 Comments and Bibliographical Remarks

This chapter is largely based on the work [26] (written together with our friend and colleague Francesco Russo). Related deterministic results can be found in the monograph [90].

In the case of bounded domain, Theorem 6.5.1 remains true for $m \in \left[\frac{d-2}{d+2}, 1\right)$ (see Theorem 3.7.3). One might suspect that also in this case the extinction property (6.77) holds for a larger class of exponents m. However, the analysis carried out in [90] for deterministic fast diffusion equations in $\mathbb{R}^d$ shows that the extinction property is dependent not only on the exponent m, but also on the space $L^p(\mathbb{R}^d)$, where the solution exists (the so called extinction space).

The analysis in this section holds in particular if all the coefficients μ_k do vanish, i.e. in the deterministic framework. In that case, Theorem 6.5.1 implies the existence of a deterministic time $\tau > 0$ so that

$$t \geq \tau \Rightarrow |X(t)| = 0,$$

and so $X(t) = 0$ for all $t \geq \tau$.

It should be said also that the exponent $m = \frac{d-2}{d+2}$ arising in Theorem 6.5.1 has a special significance because in this case for $d > 3$ Eq. (6.2) (in the deterministic case) describes the evolution of a conformal metric by the so called *Yamabe flow*.

Theorem 6.4.4 can be compared most closely to the main existence result of [86], see also [84] obtained via variational approach (Chap. 4). But there are, however, a few notable differences we explain below. The functions ψ arising in [86] are monotonically increasing, continuous and are assumed to satisfy a growth condition of the form

$$N(r) \leq r\psi(r) \leq C(N(r) + 1)r, \quad \forall r \in \mathbb{R},$$

where N is a smooth and Δ_2-regular Young function defining the Orlicz class L_N. In contrast to this, here ψ is any maximal monotone multivalued graph with arbitrary polynomial growth. In [84, 86], however, existence and uniqueness is obtained for all initial conditions $x \in \mathcal{H}^{-1}$ without any further restrictions.

In [27] a probabilistic representation for solutions to Eq. (6.2) in 1-D, is proved.

Chapter 7
Transition Semigroup

This chapter is devoted to existence of invariant measures for transition semigroups associated with stochastic porous media equations with additive noise studied in previous chapters.

7.1 Introduction and Preliminaries

We are concerned with the stochastic differential equation (3.1) under Hypothesis 4. By Theorem 3.4.1 Eq. (3.1) has a unique generalized solution $X(\cdot, x) \in L_W^2(\Omega; C([0, T]; H^{-1}))$ for every $x \in H^{-1}$.

We will study here the *transition semigroup* P_t, $t \geq 0$, defined for $\varphi \in B_b(H^{-1})$ by

$$P_t \varphi(x) := \mathbb{E}[\varphi(X(t, x))], \quad \forall\, t \geq 0,\ x \in H^{-1}. \tag{7.1}$$

and its invariant measures. For the sake of simplicity we shall limit ourselves to equations with additive noise, of the form

$$\begin{cases} dX(t) = \Delta\beta(X(t))dt + \sqrt{Q}\,dW(t), \\ X(0) = x \in H^{-1}. \end{cases} \tag{7.2}$$

Here Q is a linear positive operator in H^{-1} and Hypothesis 4 requires that for $\sigma := \sqrt{Q}$ we have $\sigma \in \mathscr{L}_2(H^{-1}, L^2)(\subset \mathscr{L}_2(H^{-1}))$, which we shall assume in the entire chapter without further notice.

So, with the notations introduced in Hypothesis 4, we have

$$\sigma_1^2 = \operatorname{Tr} Q, \quad \sigma_2^2 = \|\sigma\|_{\mathscr{L}_2(H^{-1}, L^2)}.$$

© Springer International Publishing Switzerland 2016
V. Barbu et al., *Stochastic Porous Media Equations*, Lecture Notes
in Mathematics 2163, DOI 10.1007/978-3-319-41069-2_7

Since $X(t, \cdot)$ is continuous for any $t \geq 0$, it is obvious that if $\varphi \in C_b(H^{-1})$ then $P_t\varphi$ is continuous as well, that is P_t, $t \geq 0$, is a *Feller* transition semigroup.

By (7.2) it follows that P_t is nonnegative for any nonnegative function $\varphi \in B_b(H^{-1})$. Moreover

$$P_t \mathbb{1} = \mathbb{1}, \quad \forall\, t \geq 0.$$

(By $\mathbb{1}$ we mean the function identically equal to 1.) We say that P_t, $t \geq 0$, is a *Markov transition semigroup*.

We shall use the following notations:

(i) $\mathscr{P}(H^{-1})$ is the space of all Borel probability measures on H^{-1}.
(ii) $B_b(H^{-1})$ is the space of all Borel and bounded real functions on H^{-1}.
(iii) $C_b(H^{-1})$ is the space of all real functions in H^{-1} which are uniformly continuous and bounded. Spaces $C_b^k(H^{-1})$, $k \in \mathbb{N}$, are defined similarly in the usual way.

Using formula (7.1) we can extend P_t, $t \geq 0$, to the space $B_b(H^{-1})$. The classical proof is presented in the next proposition for the reader's convenience.

Proposition 7.1.1 $P_t\varphi \in B_b(H^{-1})$ *for all* $\varphi \in B_b(H^{-1})$.

Proof Let first $\varphi = \mathbb{1}_C$, where C is a closed subset of H^{-1} and $\mathbb{1}_C$ is the characteristic function of C,

$$\mathbb{1}_C := \begin{cases} 1 & \text{if } x \in C \\ \\ 0 & \text{if } x \notin C. \end{cases}$$

It is well known that there exists a sequence $\{\varphi_n\} \subset C_b(H^{-1})$ monotonically decreasing to φ, so that $\{P_t\varphi_n\}$ is monotonically convergent to $P_t\varphi$. On the other hand, $P_t\varphi_n$ is continuous for all $n \in \mathbb{N}$ since P_t is Feller. So, $P_t\varphi$ is measurable. Next let $\varphi = \mathbb{1}_I$, where I is a Borel subset of H^{-1}. Then we conclude that $P_t\varphi$ is measurable by a monotone classes argument. $\qquad\qquad\square$

Finally, for any $t \geq 0$ and any $x \in H^{-1}$ we denote by $\pi_{t,x}$ the law of $X(t, x)$, so that

$$P_t\varphi(x) = \int_H \varphi(y)\pi_{t,x}(dy), \qquad \forall\, \varphi \in B_b(H^{-1}). \tag{7.3}$$

In particular, we have

$$\pi_{0,x} = \delta_x, \quad \forall\, x \in H^{-1}$$

and for any $\Gamma \in \mathscr{B}(H^{-1})$

$$P_t \mathbb{1}_\Gamma(x) = \pi_{t,x}. \tag{7.4}$$

Proposition 7.1.2 *The following statements hold.*

(i) *For any $I \in \mathscr{B}(H^{-1})$ the mapping*

$$[0, \infty) \times H \to \mathbb{R}, \quad (t, x) \mapsto \pi_{t,x}(I),$$

is Borel.

(ii) *For any $\Gamma \in \mathscr{B}(H^{-1})$, $t, s \geq 0$, $x \in H$, the following Chapman–Kolmogorov equation holds*

$$\pi_{t+s,x}(\Gamma) = \int_H \pi_{s,y}(\Gamma) \pi_{t,x}(dy). \tag{7.5}$$

We omit the standard proof. We say that the mapping

$$[0, +\infty) \times H^{-1} \to \mathscr{P}(H^{-1}), \quad (t, x) \mapsto \pi_{t,x}$$

is a *probability kernel*.

7.1.1 The Infinitesimal Generator of P_t

We recall that $u_n \in C_b(H^{-1})$, $n \in \mathbb{N}$, is π-convergent to $u \in C_b(H^{-1})$ if

$$\sup_{n \in \mathbb{N}} \|u_n\|_{C_b(H^{-1})} < \infty \quad \text{and} \quad \lim_{n \to \infty} u_n(x) = u(x) \quad \forall \, x \in H^{-1}.$$

Let us now introduce the infinitesimal generator $\mathscr{K}$ of the semigroup P_t on $C_b(H^{-1})$. Set

$$\Delta_\epsilon := \frac{1}{\epsilon} \left(P_\epsilon - I \right), \quad \epsilon \in (0, 1].$$

Definition 7.1.3 We say that $\varphi \in C_b(H^{-1})$ belongs to the domain $D(\mathscr{K})$ of $\mathscr{K}$ if there exists a function $\psi \in C_b(H^{-1})$ such that

(i) $\lim_{\epsilon \to 0} \dfrac{1}{\epsilon} \left(P_\epsilon(x) - x \right) = \psi(x)$ for all $x \in H^{-1}$.

(ii) $\sup_{\epsilon \in (0, 1]} \|\Delta_\epsilon \varphi\|_{C_b(H^{-1})} < \infty$.

If $\varphi \in D(\mathscr{K})$ we write $\mathscr{K}\varphi = \psi$.

We shall denote by $\rho(\mathscr{K})$ the resolvent set of $\mathscr{K}$, i.e. the set of all $\lambda \in \mathbb{R}$ such that

$$\lambda - \mathscr{K} : D(\mathscr{K}) \to C_b(H^{-1})$$

is bijective and its resolvent $R(\lambda, \mathcal{K}) := (\lambda - \mathcal{K})^{-1}$ is π-continuous, i.e. $R(\lambda, \mathcal{K})u_n$ is π-convergent to $R(\lambda, \mathcal{K})u$, if u_n is π-convergent to u.

The following proposition is a generalization, proved in [49], of a result established in [43, 83]. We omit the proof.

Proposition 7.1.4 *For any $\lambda > 0$ and any $f \in C_b(H^{-1})$ we have $\lambda \in \rho(\mathcal{K})$ and*

$$R(\lambda, \mathcal{K})f(x) = \int_0^\infty e^{-\lambda t} P_t f(x)dt, \quad x \in H^{-1}. \tag{7.6}$$

We recall now that a Borel probability measure μ in H is invariant for P_t, $t \geq 0$, if

$$\int_{H^{-1}} P_t \varphi \, d\mu = \int_{H^{-1}} \varphi \, d\mu, \quad \forall \, \varphi \in C_b(H^{-1}), \, \forall \, t > 0.$$

We are going to study existence and uniqueness of invariant measures for P_t, $t \geq 0$. We shall deal with the following situations

(i) Slow diffusions, (ii) Stefan problem, (iii) Fast diffusions, (iv) Equation from the self organized criticality.

Throughout in the following we shall use the notation

$$\kappa := \sup\{\|x\|_{-1}^2 : |x|_2^2 \leq 1\},$$

so that

$$\|x\|_{-1}^2 \leq \kappa |x|_2^2, \quad \forall \, x \in L^2. \tag{7.7}$$

7.2   Invariant Measures for the Slow Diffusions Semigroup P_t

We consider here the Eq. (7.1) with $\beta(r) = |r|^{2m} r$ with $m > 0$,

$$\begin{cases} dX(t) = \Delta(|X(t)|^{2m} X(t)) \, dt + \sqrt{Q} \, dW(t), \\ \\ X(0) = x \in H^{-1}. \end{cases} \tag{7.8}$$

We shall extend a method from in [51], which was adapted from the mild to the variational framework in [76, Theorem 4.3.9] and which does not apply here since β is not strictly monotone.[1]

---

[1] β is called *strictly monotone* if there exists $a > 0$ such that $(\beta(r) - \beta(s))(r - s) \geq a|r - s|^2$, for all $r, s \in \mathbb{R}$.

We shall show that there exists $\eta \in L^2(\Omega, \mathcal{F}, \mathbb{P}; H^{-1})$ such that

$$\lim_{\lambda \to +\infty} X(t, -\lambda, x) = \eta \quad \text{in } L^2(\Omega, \mathcal{F}, \mathbb{P}; H^{-1}),$$

which implies that the law of η is the unique invariant measure of P_t.

Here $X(t, -\lambda, x)$ is the solution of the following problem (which can be solved, with minor changes, as problem (7.2))

$$\begin{cases} dX(t) = \Delta\beta(X(t))dt + \sqrt{Q}\, d\overline{W}(t) \\ X(-\lambda) = x, \end{cases} \tag{7.9}$$

where

$$\overline{W}(t) = \begin{cases} W(t), & \text{if } t \geq 0, \\ W_1(-t), & \text{if } t \leq 0, \end{cases}$$

and W_1 is another H^{-1}-valued Brownian motion independent of W.

We shall use the following elementary fact. There is $c_m > 0$ such that

$$2(\beta(r) - \beta(s))(r - s) \geq c_m(r - s)^{2m+2}, \quad \forall\, r, s \in \mathbb{R}. \tag{7.10}$$

Remark 7.2.1 A different proof of existence and uniqueness of the invariant measure as well as of estimate (7.12) below is given in [53].   □

Theorem 7.2.2 *Assume that* $\beta(r) = |r|^{2m}r$ *with* $m > 0$. *Let* $x \in H^{-1}$, $t \geq 0$ *and let* X *be the solution of* (7.8). *Then there exists the limit*

$$\lim_{\lambda \to +\infty} X(t, -\lambda, x) = \eta \quad \text{in } L^2(\Omega, \mathcal{F}, \mathbb{P}; H^{-1}). \tag{7.11}$$

Moreover there exists $K_m > 0$ *such that*

$$\mathbb{E}\|X(t, -\lambda, x) - \eta\|_{-1} \leq K_m(t + \lambda)^{-\frac{1}{m}}, \quad t > 0. \tag{7.12}$$

for some $c > 0$ *and any* $\lambda > 0$, $t \geq -\lambda$.

Finally, the law μ *of* η *is the unique invariant measure of* P_t, $t \geq 0$, *and we have*

$$\lim_{t \to \infty} P_t\varphi(x) = \int_{H^{-1}} \varphi(y)\mu(dy), \quad \forall\, \varphi \in C_b(H). \tag{7.13}$$

Therefore μ *is ergodic and strongly mixing (see e.g.* [50]).

Proof Set

$$X_\lambda(t) = X(t, -\lambda, x), \quad \forall\, t \geq -\lambda$$

and for $\lambda > \epsilon > 0$,

$$Z(t) = X_\lambda(t) - X_\epsilon(t), \quad \forall\, t \geq -\epsilon.$$

Then, if $x \in L^2$,

$$dZ(t) = (\Delta\beta(X_\lambda(t)) - \Delta\beta(X_\epsilon(t)))dt, \quad t \geq -\epsilon.$$

It follows that

$$\frac{d}{dt}\,\|Z(t)\|_{-1}^2 = 2\langle Z(t), \Delta\beta(X_\lambda(t)) - \Delta\beta(X_\epsilon(t))\rangle_{-1}$$

$$= -2\langle Z(t), \beta(X_\lambda(t)) - \beta(X_\epsilon(t))\rangle_2.$$

Integrating with respect to t in $[-\epsilon, t]$ and taking expectation, yields

$$\mathbb{E}\|Z(t)\|_{-1}^2 = -2\mathbb{E}\int_{-\epsilon}^{t} \langle Z(s), \beta(X_\lambda(s)) - \beta(X_\epsilon(s))\rangle_2\, dt.$$

Therefore, for $t \geq -\epsilon$

$$\frac{d}{dt}\,\mathbb{E}\|Z(t)\|_{-1}^2 = -2\mathbb{E}\langle Z(t), \beta(X_\lambda(t)) - \beta(X_\epsilon(t))\rangle_2. \tag{7.14}$$

Now by (7.10) we have

$$2\mathbb{E}\langle Z(t), \beta(X_\lambda(t)) - \beta(X_\epsilon(t))\rangle_2 \geq c_m |Z(t)|_{2m+2}^{2m+2},$$

so that

$$\frac{d}{dt}\,\mathbb{E}\|Z(t)\|_{-1}^2 \leq -c_m \mathbb{E}|Z(t)|_{2m+2}^{2m+2}. \tag{7.15}$$

Since $L^{2m+2} \subset H^{-1}$ with continuous embedding, there exists $c_{1,m} > 0$ such that

$$\frac{d}{dt}\,\mathbb{E}\|Z(t)\|_{-1}^2 \leq -c_{1,m}\, \mathbb{E}\|Z(t)\|_{-1}^{2m+2} \leq -c_{1,m}\, (\mathbb{E}\|Z(t)\|_{-1}^2)^{m+1}. \tag{7.16}$$

(In the second inequality we have used the Hölder inequality.) Now, by a standard comparison result we find

$$\mathbb{E}\|Z(t)\|_{-1}^2 \leq \frac{\mathbb{E}\|Z(-\epsilon)\|_{-1}^2}{(1 + c_{1,m}(t+\epsilon)(\mathbb{E}\|Z(-\epsilon)\|_1^2)^{m+1})^{1/m}}, \quad \forall\, t \geq -\epsilon.$$

Consequently

$$\mathbb{E}\|Z(t)\|_{-1}^2 \leq \frac{1}{[c_{1,m}(t+\epsilon)]^{1/m}}, \quad \forall\, t \geq -\epsilon.$$

In conclusion, setting $K_m := c_{1,m}^{-1/m}$, we have

$$\mathbb{E}\|X(t,-\lambda,x) - X(t,-\epsilon,x)\|_{H^{-1}}^2 \leq \frac{K_m}{(t+\epsilon)^{1/m}}, \quad \forall\, t \geq -\epsilon. \tag{7.17}$$

By density this extends to generalized solutions, hence to all $x \in H^{-1}$.

It follows that $\{X(t,-\lambda,x)_{\lambda>0}\}$ is a Cauchy sequence in $L^2(\Omega,\mathscr{F},\mathbb{P};H^{-1})$ so that there exists the limit

$$\lim_{\lambda\to\infty} X(t,-\lambda,x) =: \eta \quad \text{in } L^2(\Omega,\mathscr{F},\mathbb{P};H^{-1}).$$

Moreover, letting $\lambda \to \infty$ in (7.17), yields

$$\mathbb{E}\|\eta - X(t,-\epsilon,x)\|_{H^{-1}}^2 \leq \frac{K_m}{(t+\epsilon)^{1/m}}, \quad t \geq -\epsilon, \tag{7.18}$$

which proves (7.12).

Let us now denote by μ the law of η. Take $\varphi \in C_b(H^{-1})$, then letting $t \to +\infty$ in the identity

$$P_t\varphi(x) = \mathbb{E}[\varphi(X(t,x))] = \mathbb{E}[\varphi(X(0,-t,x))],$$

yields

$$\lim_{t\to+\infty} P_t\varphi(x) = \mathbb{E}[\varphi(\eta)] = \int_{H^{-1}} \varphi(y)\mu(dy), \quad \forall\, x \in H^{-1}. \tag{7.19}$$

Now we can show the invariance of μ. Write for $\varphi \in C_b(H^{-1})$, $t, s > 0$,

$$\int_{H^{-1}} P_{t+s}\varphi\, d\mu = \int_{H^{-1}} P_t P_s \varphi\, d\mu.$$

Letting $t \to +\infty$ yields

$$\int_{H^{-1}} \varphi\, d\mu = \int_{H^{-1}} P_s \varphi\, d\mu,$$

so that μ is invariant and (7.19) is proved. Let us prove uniqueness. Let ν be another invariant measure, so that

$$\int_{H^{-1}} P_t\varphi(x)\nu(dx) = \int_{H^{-1}} \varphi(x)\nu(dx), \quad \forall \, \varphi \in C_b(H^{-1}). \tag{7.20}$$

Then, letting $t \to +\infty$, we find by (7.19) that

$$\int_{H^{-1}} \varphi(x)\mu(dx) = \int_{H^{-1}} \varphi(x)\nu(dx),$$

so that $\nu = \mu$. The proof is complete. $\square$

Remark 7.2.3 Let η be defined by (7.11). It is easy to see that η is independent of x. Obviously, $\eta = \eta_t$ depends on t in general and $t \to \eta_t$ is the stationary solution to problem (7.2). $\square$

Remark 7.2.4 By Theorem 7.2.2 the unique invariant measure is ergodic and strongly mixing. Therefore by the classical Birkhoff–Von Neumann theorem we have

$$\lim_{t\to\infty} P_t(\mathbb{1}_\Gamma)(x) = \lim_{t\to\infty} \mathbb{P}(X(t,x) \in \Gamma) = \mu(\Gamma),$$

for all $x \in H^{-1}$ and any Borel subset Γ of H^{-1}.

$\square$

Proposition 7.2.5 *Let μ be the invariant measure of P_t, $t \geq 0$. Then we have*

$$\int_{H^{-1}} \|x\|^2_{-1}\mu(dx) < \infty \tag{7.21}$$

$$\int_{H^{-1}} \langle \beta(x), x\rangle_2\, \mu(dx) = \int_{H^{-1}} |x|^{2m+2}_{2m+2} \mu(dx) = Tr\, Q. \tag{7.22}$$

Consequently, the support of μ is included in L^{2m+2}.

Proof We first notice that

$$\int_{H^{-1}} \|x\|^2_{-1}\mu(dx) = \mathbb{E}[\|\eta\|^2_{-1}],$$

so that (7.21) is fulfilled.

Now integrating with respect to μ over H^{-1} the identity

$$\mathbb{E}\|X(t,x)\|^2_{-1} + \mathbb{E}\int_0^t \langle \beta(X(s,x)), X(s,x)\rangle_2\, ds = \|x\|^2_{-1} + t\, Tr\, Q,$$

and dividing by t and letting $t \to \infty$ we get the desired result. $\square$

7.3 Invariant Measure for the Stefan Problem

We consider here the Eq. (7.1) where β is given by (1.14)

$$\beta(r) = \begin{cases} \alpha_1 r & \text{for } r < 0, \\ 0 & \text{for } 0 \leq r \leq \rho, \\ \alpha_2(r - \rho) & \text{for } r > \rho, \end{cases} \tag{7.23}$$

$\alpha_1, \alpha_2, \rho > 0$. Note that

$$\frac{r\beta(r)}{r^2} = \begin{cases} \alpha_1, & \text{if } r < 0, \\ 0, & \text{if } 0 \leq r < \rho \\ \alpha_2 \frac{r-\rho}{r}. & \text{if } r > \rho. \end{cases}$$

Therefore

$$\lim_{r \to -\infty} \frac{r\beta(r)}{r^2} = \alpha_1, \qquad \lim_{r \to +\infty} \frac{r\beta(r)}{r^2} = \alpha_2.$$

Consequently, setting

$$c_1 = \min\{\alpha_1, \alpha_2\},$$

there exists $K > 0$ such that

$$2r\beta(r) \geq c_1 r^2, \quad \forall \, r > K.$$

Now, setting

$$g := -\inf\{2r\beta(r) : \ |r| \leq K\},$$

we have

$$2r\beta(r) \geq c_1 r^2 - g, \quad \forall \, r \in \mathbb{R}. \tag{7.24}$$

Theorem 7.3.1 *There exists an invariant measures μ for P_t, $t \geq 0$, whose support is included in L^2. Moreover*

$$\int_{H^{-1}} \|x\|_{-1}^2 \, \mu(dx) \leq \frac{\kappa(g + Tr\, Q)}{c_1}. \tag{7.25}$$

and

$$\int_{H^{-1}} |x|_2^2 \, \mu(dx) \leq \frac{1}{c_1} \, (g + Tr\, Q). \tag{7.26}$$

Proof Let $x \in H^{-1}$. Then by Itô's formula and standard computations we have

$$\mathbb{E}\|X(t,x)\|_{-1}^2 + 2\mathbb{E} \int_0^t \langle X(s,x), \beta(X(s,x))\rangle_2 \, ds = \|x\|_{-1}^2 + t \, \mathrm{Tr} \, Q. \qquad (7.27)$$

Taking into account (7.24) it follows that

$$\mathbb{E}\|X(t,x)\|_{-1}^2 + c_1 \int_0^t \mathbb{E}|X(s,x)|_2^2 \, ds \leq \|x\|_{-1}^2 + t(g + \mathrm{Tr} \, Q), \qquad (7.28)$$

which implies

$$c_1 \int_0^t \mathbb{E}|X(s,x)|_2^2 \, ds \leq \|x\|_{-1}^2 + (g + \mathrm{Tr} \, Q)t, \quad t > 0. \qquad (7.29)$$

Now we fix $x \in H^{-1}$ and set

$$\mu_t = \frac{1}{t} \int_0^t \pi_{s,x} ds, \quad \forall \, t \geq 1.$$

(We recall that $\pi_{s,x}$ is the law of $X(s,x)$.) We are going to show that the family of measures $\{\mu_t\}_{t \geq 1}$ is tight in H^{-1}. Let $R > 0$, and let B_R denote the open ball of L^2 of center 0 and radius R and B_R^c its complement in L^2. Then for any $t \geq 1$ we have

$$\pi_{t,x}(B_R^c) = \int_{|y|_2 \geq R} \pi_{t,x}(dy)$$

$$\leq \frac{1}{R^2} \int_{H^{-1}} |y|_2^2 \, \pi_{t,x}(dy) = \frac{1}{R^2} \mathbb{E}|X(t,x)|_2^2.$$

Taking into account (7.29) we conclude that

$$\mu_t(B_R^c) \leq \frac{1}{tc_1 R^2} \left(\|x\|_{-1}^2 + (g + \mathrm{Tr} \, Q)t \right).$$

Therefore

$$\mu_t(B_R^c) \leq \frac{\|x\|_{-1}^2 + g + \mathrm{Tr} \, Q}{c_1 R^2}, \quad \forall \, t \geq 1.$$

Since the imbedding $L^2 \subset H^{-1}$ is compact this implies that $\{\mu_t\}_{t \geq 1}$ is tight in H^{-1}. Now the existence of an invariant measure follows from the Krylov–Bogoliubov theorem.

Let us show (7.26) and (7.25) . We first notice that (7.28) can be written as

$$\int_{H^{-1}} \|y\|_{-1}^2 \, \pi_{t,x}(dy) + c_1 \int_0^t \int_{H^{-1}} |y|_2^2 \, \pi_{s,x}(dy) ds$$

$$\leq (\|x\|_{-1}^2 + t(g + \mathrm{Tr}\, Q))$$

It follows that for any $L > 0$

$$\frac{1}{t} \int_{H^{-1}} \frac{\|y\|_{-1}^2}{1 + L\|y\|_{-1}^2} \, \pi_{t,x}(dy) + \frac{c_1}{t} \int_0^t \int_{H^{-1}} \frac{|y|_2^2}{1 + L|y|_2^2} \, \pi_{s,x}(dy) ds$$

$$\leq \frac{1}{t} \|x\|_{-1}^2 + g + \mathrm{Tr}\, Q.$$

Letting $t \to \infty$ and recalling that the invariant measure μ is a weak limit point of $\{\mu_t\}_{t>1}$ and that its support is included in L^2, yields

$$\int_{H^{-1}} \frac{|y|_2^2}{1 + L|y|_2^2} \, \mu(dy) \leq g + \mathrm{Tr}\, Q.$$

Now (7.25) follows letting $L \to \infty$.

Finally, (7.26) follows from (7.25) recalling (7.7). $\qquad\square$

Remark 7.3.2 Let Σ be the set of all invariant measures of P_t, $t \geq 0$ with finite second moment. We know by Theorem 7.3.1 that Σ is non empty. Now let $\nu \in \Sigma$. Then integrating both sides of (7.29) with respect to ν and letting $t \to \infty$ yields

$$\int_{H^{-1}} |x|_2^2 \, \nu(dx) \leq \frac{1}{c_1} (K_1 + g). \qquad (7.30)$$

This implies that Σ is tight; since it is weakly closed it follows that it is weakly compact. Then by the Krein–Milman theorem Σ contains an extremal point, which is an ergodic measure, see e.g. [50]. $\qquad\square$

Remark 7.3.3 The uniqueness of invariant measure for the Stefan problem is an open problem. However, uniqueness happens for the parabolic phase field system

$$\begin{cases} du + \ell d\varphi - k\Delta u dt = \sqrt{Q_1} \, dW_1 \\ d\varphi - \alpha \Delta\varphi - \delta(\varphi - \varphi^3) - \gamma u = \sqrt{Q_2} \, dW_2, \end{cases}$$

where $\alpha, \delta, \gamma, k, \ell$ are positive constants and W_1, W_2 are independent cylindrical Wiener processes in L^2, which replaces the two-phase Stefan problem as model of phase transition (see [12]). $\qquad\square$

7.4   Invariant Measures for Fast Diffusions

We consider here problem (7.2) with $\beta(r) = |r|^{\alpha-1}r$, $\alpha \in (0, 1)$, i.e.,

$$
\begin{cases}
dX(t) = \Delta(|X(t)|^{\alpha-1}X(t))dt + \sqrt{Q}\,dW(t), \\[2mm]
X(0) = x \in H^{-1}.
\end{cases}
\tag{7.31}
$$

We are going to show existence of an invariant measure for the transition semigroup P_t, $t \geq 0$. Uniqueness will be proved when $d = 1, 2$ or $d < \frac{1+\alpha}{1-\alpha}$.

7.4.1   Existence

We first need an inequality which follows from (3.22) . For all $m \in \mathbb{N}$ there exists $C > 0$ (depending on m) such that

$$
\mathbb{E}|X(t)|_{2m}^{2m} + 2m\alpha(2m-1))\int_0^t \mathbb{E}[|X(s)|^{2m+\alpha-3}\,|\nabla X(s)|^2]ds
\tag{7.32}
$$

$$
\leq |x|_{2m}^{2m} + Ct.
$$

Since obviously

$$
|X(s)|^{2m+\alpha-3}\,|\nabla X(s)|^2 = \frac{4\alpha}{2m+\alpha-1}\nabla\left(|X(s)|^{(2m+\alpha-1)/2}\right),
$$

we deduce from (7.32) that

$$
\mathbb{E}|X(t)|_{2m}^{2m} + \frac{8\alpha^2 m}{(2m+\alpha-1)(2m-1)}\int_0^t \mathbb{E}|\nabla(|X(s)|^{(2m+\alpha-1)/2})|^2\,ds
\tag{7.33}
$$

$$
\leq |x|_{2m}^{2m} + Ct.
$$

By the Sobolev embedding theorem there exists a constant $C_1 > 0$ (depending on m) such that

$$
\mathbb{E}|X(t)|_{2m}^{2m} + C_1\mathbb{E}\int_0^t |X(s)|_{p(m)}^{p(m)}ds \leq |x|_{2m}^{2m} + Ct,
\tag{7.34}
$$

where

$$p(m) = \begin{cases} \in [0, \infty) & \text{if } d = 1, 2, \\ = \frac{(2m+\alpha-1)d}{d-2} & \text{if } d > 2. \end{cases} \tag{7.35}$$

We can now show the existence of an invariant measure.

Theorem 7.4.1 *Assume that $\beta(r) = |r|^{\alpha-1}r$, $\alpha \in (0, 1)$. Then there exists an invariant measure μ for P_t, $t \geq 0$. Moreover, we have*

$$\int_{H^{-1}} |x|_p^p \, \mu(dx) < \infty, \tag{7.36}$$

for all $p \geq 1$.

Proof Step 1. Existence.

Let $x \in L^{2m}$, $m > 1$. Then by (7.34) we have

$$\mathbb{E}|X(t)|_{2m}^{2m} + C_m \mathbb{E} \int_0^t |X(s)|_{p(m)}^{p(m)} ds \leq |x|_{2m}^{2m} + Ct, \tag{7.37}$$

Therefore, there exists $C_2 > 0$ such that

$$\frac{1}{t} \int_0^t \mathbb{E}|X(s,x)|_{p(m)}^{p(m)} \, ds \leq \frac{1}{t} |x|_{2m}^{2m} + C_2, \quad t > 0. \tag{7.38}$$

Now set

$$\mu_t = \frac{1}{t} \int_0^t \pi_{s,x} ds,$$

(recall that $\pi_{s,x}$ is the law of $X(s, x)$.)

We are going to show that $\{\mu_t\}_{t \geq 1}$ is tight. Let $R > 0$, and let B_R denote the ball of $L^{p(m)}$ of center 0 and radius R and B_R^c its complement in $L^{p(m)}$. Then for any $t \geq 1$ we have

$$\pi_{t,x}(B_R^c) = \int_{|y|_{p*} \geq R} \pi_{t,x}(dy)$$

$$\leq \frac{1}{R^{p*}} \int_{H^{-1}} |y|_{p(m)}^{p(m)} \pi_{t,x}(dy) = \frac{1}{R^{p*}} \mathbb{E}|X(t,x)|_{p(m)}^{p(m)}.$$

Taking into account (7.38) we conclude that

$$\mu_t(B_R^c) \le \frac{|x|_{p(m)}^{p(m)} + C_2}{R^2}, \quad \forall\, t \ge 1.$$

Choosing m sufficiently large we see that the imbedding $L^{p(m)} \subset H^{-1}$ is compact this implies that $\{\mu_t\}_{t \ge 1}$ is tight. Now the existence of an invariant measure follows from the Krylov–Bogoliubov theorem.

Step 2. Proof of (7.36).

First we notice that we can write (7.37) as

$$\int_{H^{-1}} |y|_{2m}^{2m} \pi_{t,x}(dy) + C_m \int_0^t \int_{H^{-1}} |y|_{p^*}^{p^*} \pi_{s,x}(dy)\, ds$$

$$\le |x|_{2m}^{2m} + Ct,$$

It follows that for any $L > 0$

$$\frac{1}{t} \int_{H^{-1}} \frac{|y|_{2m}^{2m}}{1 + L|y|_{2m}^{2m}} \pi_{t,x}(dy) + \frac{1}{t} C_m \int_0^t \int_{H^{-1}} \frac{|y|_{p^*}^{p^*}}{1 + L|y|_{p(m)}^{p(m)}} \pi_{s,x}(dy)\, ds$$

$$\le \frac{1}{t} |x|_{2m}^{2m} + C.$$

Letting $t \to \infty$ and recalling that μ is a weak limit point of $\{\mu_t\}_{t>1}$ and that its support is included in $L^{p(m)}$, yields

$$\int_{H^{-1}} \frac{|y|_{p(m)}^{p(m)}}{1 + L|y|_{p(m)}^{p(m)}} \mu(dy) \le \frac{C}{C_m}.$$

Now (7.36) follows letting $L \to \infty$ and taking into account the arbitrariness of m.

$\square$

7.4.2 Uniqueness

We start with an elementary lemma due to [77].

Lemma 7.4.2 *For any $\alpha \in (0,1)$, $r, s \in \mathbb{R}$ we have*

$$(\beta(r) - \beta(s))(r - s) \ge \alpha \frac{(r - s)^2}{(|r| \vee |s|)^{1-\alpha}}. \tag{7.39}$$

Proof It is enough to show (7.39) for $r > s > 0$ and $\alpha = \frac{1}{n}$ with $n \in \mathbb{N}$, $n \geq 2$. In this case (7.39) reduces to

$$(r - s)(r^{\frac{1}{n}} - s^{\frac{1}{n}}) \geq \frac{1}{n} \frac{(r - s)^2}{r^{1 - \frac{1}{n}}}. \tag{7.40}$$

Setting $r^{\frac{1}{n}} = u$ and $s^{\frac{1}{n}} = v$, (7.40) reduces to

$$(u^n - v^n)(u - v) \geq \frac{1}{n} \frac{(u^n - v^n)^2}{u^{n-1}}, \tag{7.41}$$

which yields

$$nu^{n-1} \geq \frac{u^n - v^n}{u - v},$$

equivalently

$$nu^{n-1} \geq u^{n-1} + u^{n-2}v + \cdots + v^{n-1} \tag{7.42}$$

This is obvious since $u \geq v$. □

We now need two more lemmas, also due to [77]. The first is a simple consequence of Hölder's inequality

Lemma 7.4.3 *Let* $(Z, \mathscr{F}, \mu)$ *be a measure space and let* f, g *nonnegative and measurable real functions. Then for any* $\alpha \in (0, 1)$ *we have*

$$\left(\int_Z f^{1+\alpha} d\mu \right)^{\frac{2}{1+\alpha}} \leq \int_Z f^2 g^{\alpha-1} d\mu \left(\int_Z g^{1+\alpha} d\mu \right)^{\frac{1-\alpha}{1+\alpha}} \tag{7.43}$$

Proof Writing

$$\int_Z f^{1+\alpha} d\mu = \int_Z (f^{1+\alpha} g^{-\alpha}) g^\alpha d\mu$$

and using Hölder's inequality with exponents $p = \frac{2}{1+\alpha}$ and $q = \frac{2}{1-\alpha}$, yields

$$\int_Z f^{1+\alpha} d\mu \leq \left(\int_Z f^2 g^{-\frac{2\alpha}{1+\alpha}} d\mu \right)^{\frac{1+\alpha}{2}} \left(\int_Z g^{\frac{2\alpha}{1-\alpha}} d\mu \right)^{\frac{1-\alpha}{2}}$$

and the conclusion follows. □

We now recall that by the Sobolev embedding theorem we have $H_0^1 \subset L^\infty$ if $d = 1$, $H_0^1 \subset L^{\frac{2d}{d-2}}$ if $d > 2$ and $L^2 \subset H^{-1} \; \forall \; q > 1$ if $d = 2$. Consequently we have $L^1 \subset H^{-1}$ if $d = 1$, $L^{\frac{2d}{d+2}} \subset H^{-1}$ if $d > 2$ and $H_0^1 \subset \bigcup_{1 \le p < \infty} L^p$ if $d = 2$.

Lemma 7.4.4 *Assume that either $d = 1, 2$ or $d \le \frac{1+\alpha}{1-\alpha}$. Then there is $\delta_\alpha > 0$ such that for any $x, y \in L^{\alpha+1}$ we have*

$$2\langle \Delta\beta(x) - \Delta\beta(y), x - y \rangle_{-1} \le -\delta_\alpha \frac{\|x - y\|_{-1}^2}{|x|_{\alpha+1}^{1-\alpha} + |y|_{\alpha+1}^{1-\alpha}} \tag{7.44}$$

Proof We have

$$J := 2\langle \Delta\beta(x) - \Delta\beta(y), x - y \rangle_{-1} = -2 \int_{\mathscr{O}} (\beta(x) - \beta(y))(x - y) d\xi.$$

Taking into account (7.39), yields

$$J \le -\alpha \int_{\mathscr{O}} \frac{(x - y)^2}{(|x| + |y|)^{1-\alpha}} d\xi. \tag{7.45}$$

Now we use (7.43) with

$$f = |x - y|, \quad g = |x| + |y|$$

and obtain

$$\left(\int_{\mathscr{O}} |x - y|^{1+\alpha} d\xi \right)^{\frac{2}{1+\alpha}} \le \int_{\mathscr{O}} (x - y)^2 (|x| + |y|)^{\alpha-1} d\xi \left(\int_{\mathscr{O}} (|x| + |y|)^{1+\alpha} d\xi \right)^{\frac{1-\alpha}{1+\alpha}}$$

So, from (7.45) we deduce

$$J \le -\alpha \frac{\left(\int_{\mathscr{O}} |x - y|^{1+\alpha} d\xi \right)^{\frac{2}{1+\alpha}}}{\left(\int_{\mathscr{O}} (|x| + |y|)^{1+\alpha} d\xi \right)^{\frac{1-\alpha}{1+\alpha}}} \tag{7.46}$$

Under our assumptions we have $L^{\alpha+1} \subset H^{-1}$, with continuous embedding, by the Sobolev embedding theorem. Therefore there exists a constant $c_\alpha > 0$ such that

$$\|x\|_{-1} \le c_\alpha |x|_{\alpha+1}, \quad \forall \, x \in L^{\alpha+1}.$$

Now the conclusion follows from (7.46).   □

Lemma 7.4.5 *Assume that either $d = 1, 2$ or $d \leq \frac{1+\alpha}{1-\alpha}$. Let $x, y \in H^{-1}$. Then there is $\delta_\alpha > 0$ such that*

$$\|X(t, x) - X(t, y)\|^2_{-1}$$

$$\leq \frac{1}{\delta_\alpha t} \|x - y\|^2_{-1} \frac{1}{t} \int_0^t (|X(s, x)|^{1-\alpha}_{1+\alpha} + |X(s, y)|^{1-\alpha}_{1+\alpha}) ds. \tag{7.47}$$

Proof Write

$$\frac{d}{dt} \|X(t, x) - X(t, y)\|^2_{-1}$$

$$= -2 \int_{\mathcal{O}} (\beta(X(t, x)) - \beta(X(t, y)))(X(t, x) - X(t, y)) \, d\xi. \tag{7.48}$$

Since the right hand side is negative we deduce that $\|X(t) - Y(t)\|_{-1}$ is decreasing. Now, by (7.48) and Lemma 7.4.4 we deduce

$$\frac{d}{dt} \|X(t, x) - X(t, y)\|^2_{-1} \leq -\delta_\alpha \frac{\|X(t, x) - X(t, y)\|^2_{-1}}{|X(t, x)|^{1-\alpha}_{\alpha+1} + |X(t, y)|^{1-\alpha}_{\alpha+1}}. \tag{7.49}$$

Therefore

$$\|X(t, x) - X(t, y)\|^2_{-1} \leq \|x - y\|^2_{-1} - \delta_\alpha \int_0^t \frac{\|X(s, x) - X(s, y)\|^2_{-1}}{|X(s, x)|^{1-\alpha}_{\alpha+1} + |X(s, y)|^{1-\alpha}_{\alpha+1}} ds. \tag{7.50}$$

Now, taking into account that $\|X(t, x) - X(t, y)\|_{-1}$ is decreasing in t, we can write

$$\|X(t, x) - X(t, y)\|^2_{-1} \leq \|x - y\|^2_{-1}$$

$$-\delta_\alpha \|X(t, x) - X(t, y)\|^2_{-1} \int_0^t \frac{1}{|X(s, x)|^{1-\alpha}_{\alpha+1} + |X(s, y)|^{1-\alpha}_{\alpha+1}} ds, \tag{7.51}$$

which implies

$$\|X(t, x) - X(t, y)\|^2_{-1} \leq \frac{\|x - y\|^2_{-1}}{1 + \delta_\alpha \int_0^t \frac{1}{|X(s, x)|^{1-\alpha}_{\alpha+1} + |X(s, y)|^{1-\alpha}_{\alpha+1}} ds} \tag{7.52}$$

Now we use the Jensen inequality[2] which implies

$$\|X(t,x) - X(t,y)\|_{-1}^2 \leq \frac{\|x-y\|_{-1}^2}{1 + \delta_\alpha \dfrac{\dfrac{t^2}{\displaystyle\int_0^t |X(s,x)|_{\alpha+1}^{1-\alpha} + |X(s,y)|_{\alpha+1}^{1-\alpha}}}{}} ds$$

that is

$$\|X(t,x) - X(t,y)\|_{-1}^2 \leq \frac{\|x-y\|_{-1}^2 \displaystyle\int_0^t (|X(s,x)|_{1+\alpha}^{1-\alpha} + |X(s,y)|_{1+\alpha}^{1-\alpha} ds)}{\delta_\alpha t^2 + \displaystyle\int_0^t (|X(s,x)|_{1+\alpha+1}^{1-\alpha} + |X(s,y)|_{1+\alpha}^{1-\alpha}) ds},$$

and a fortiori

$$\|X(t,x) - X(t,y)\|_{-1}^2 \leq \frac{1}{\delta_\alpha t} \|x-y\|_{-1}^2 \frac{1}{t} \int_0^t (|X(s,x)|_{1+\alpha}^{1-\alpha} + |X(s,y)|_{1+\alpha}^{1-\alpha}) ds.$$

$\square$

We are now ready to show

Theorem 7.4.6 *Assume that either* $d = 1, 2$ *or* $d \leq \frac{1+\alpha}{1-\alpha}$. *Then there is a unique invariant measure for* P_t, $t \geq 0$.

Proof By Itô's formula we have

$$\mathbb{E}\|X(t,x)\|_{-1}^2 + 2\mathbb{E}\int_0^t |X(s,x)|_{1+\alpha}^{1+\alpha} \, ds = \|x\|_{-1}^2 + t\sigma_1^2$$

Therefore

$$\frac{2}{t}\mathbb{E}\int_0^t |X(s,x)|_{1+\alpha}^{1+\alpha} \, ds \leq \frac{1}{t} \|x\|_{-1}^2 + \sigma_1^2, \quad \forall \, t > 0.$$

Then there is $N_\alpha > 0$ such that

$$\frac{1}{t}\mathbb{E}\int_0^t |X(s,x)|_{1+\alpha}^{1+\alpha} \, ds \leq N_\alpha (1 + \|x\|_{-1}^2 + \sigma_1^2), \quad \forall \, t \geq 1.$$

---

[2] $\frac{1}{\frac{1}{t}\int_0^t h \, dt} \leq \frac{1}{t}\int_0^t \frac{1}{h} \, dt.$

Now from (7.47) we deduce

$$\mathbb{E}\|X(t,x) - X(t,y)\|_{-1}^2 \le \frac{2N_\alpha}{\delta_\alpha t}\|x-y\|_{-1}^2(1 + \|x\|_{-1}^2 + \|y\|_{-1}^2 + \sigma_1^2), \quad \forall\, t \ge 1.$$
(7.53)

Therefore there exists $C(x,y)$ such that

$$\mathbb{E}\|X(t,x) - X(t,y)\|_{-1} \le t^{-1/2}C(x,y), \quad \forall\, t \ge 0.$$
(7.54)

It is enough to show uniqueness within the class of ergodic invariant measures (see e.g. [50]). Assume by contradiction that there exist two ergodic invariant measures μ and ν in H^{-1}. Then there exist two Borel disjoint sets A, B such that $H^{-1} = A \cup B$ and such that for any $\varphi \in C_b^1(H^{-1})$ we have

$$\begin{cases} \lim_{T\to+\infty} \frac{1}{T}\int_0^T P_t\varphi(x)dt = \int_{H^{-1}} \varphi\, d\mu, & \forall\, x \in A, \\[2mm] \lim_{T\to+\infty} \frac{1}{T}\int_0^T P_t\varphi(y)dt = \int_{H^{-1}} \varphi\, d\nu, & \forall\, y \in B. \end{cases}$$
(7.55)

Now for all $x, y \in H^{-1}$ we have

$$\left|\frac{1}{T}\int_0^T P_t\varphi(x)dt - \frac{1}{T}\int_0^T P_t\varphi(y)dt\right| \le \frac{1}{T}\int_0^T |P_t\varphi(x) - P_t\varphi(y)|dt$$

$$\le \|\varphi\|_1 \frac{1}{T}\int_0^T \mathbb{E}\|X(t,x) - X(t,y)\|_{-1}\, dt.$$

Then by (7.54) it follows that

$$\left|\frac{1}{T}\int_0^T P_t\varphi(x)dt - \frac{1}{T}\int_0^T P_t\varphi(y)dt\right| \le C(x,y)\|\varphi\|_1 \frac{1}{T}\int_0^T t^{-1/2}\, dt$$

$$= C(x,y)\|\varphi\|_1 T^{-1/2}.$$

Letting $T \to \infty$ we get

$$\lim_{T\to+\infty} \frac{1}{T}\int_0^T P_t\varphi(x)dt = \lim_{T\to+\infty} \frac{1}{T}\int_0^T P_t\varphi(y)dt, \quad \forall\, x,y \in H^{-1},$$

which is a contradiction by (7.55) □

7.5 Invariant Measure for Self Organized Criticality Equation

We consider here the Eq. (7.1) with $\beta(r)r \geq a|r| + b$, $\forall\, r \in \mathbb{R}$. In particular these assumptions are fulfilled by the self organized criticality stochastic equation (3.63) with additive noise.

$$
\begin{cases}
dX(t) = \rho\Delta(\mathrm{sign}\, X(t))dt + \sqrt{Q}\,dW(t), \\
X(0) = x \in H^{-1}.
\end{cases}
\tag{7.56}
$$

Theorem 7.5.1 *Under the assumptions above, if $d = 1$ there exists an invariant measure for P_t, $t \geq 0$. Moreover there exists $\kappa > 0$ finite such that*

$$
\int_{H^{-1}} |x|_1 \mu(dx) \leq \kappa.
\tag{7.57}
$$

Proof Arguing as in the proof of Lemma 3.7.2 we get for the solution to (3.1) the estimate

$$
\mathbb{E}\int_0^t |X(s)|_1 ds \leq \frac{1}{\rho}\,(\|x\|_{-1}^2 + Ct), \quad \forall\, t \geq 0.
\tag{7.58}
$$

Now we claim that the family of measures

$$
\left\{ \mu_t = \frac{1}{t}\int_0^t \pi_{s,x} ds \right\}_{t \geq 1}
$$

is tight in H^{-1} and so, existence of an invariant measure will follow from the Krylov–Bogoliubov theorem.

To prove the claim fix $x \in H^{-1}$ and for any $R > 0$ consider the ball

$$
B_R = \{x \in L^1 : |x|_1 < R\}.
$$

Then, denoting by B_R^c the complement of B_R in L^1, write

$$
\mu_t(B_R^c) = \frac{1}{t}\int_0^t \pi_{s,x}(B_R^c)ds
$$

$$
= \frac{1}{t}\int_0^t ds \int_{|y|_1 \geq R} \pi_{s,x}(dy)
$$

$$
\leq \frac{1}{tR}\int_0^t ds \int_{L^1} |y|_1 \pi_{s,x}(dy) = \frac{1}{tR}\int_0^t \mathbb{E}|X(s,x)|ds.
$$

Therefore, by (7.58) we deduce

$$\mu_t(B_R^c) \le \frac{1}{agR}(\|x\|_{-1} + C).$$

Taking into account that if $d = 1$ by the Sobolev embedding theorem we have $L^1 \subset H^{-1}$ with compact injection, we conclude that $(\mu_t)_{t \ge 1}$ is tight.

It remains to show (7.57). For this we write (7.58) as

$$\int_0^t \int_{H^{-1}} |y|_1 \pi_{s,x}(dy)ds \le \frac{1}{\rho}(\|x\|_{-1}^2 + Ct), \quad \forall\, t \ge 0.$$

It follows that for any $L > 0$

$$\frac{1}{t}\mathbb{E}\int_0^t \int_{H^{-1}} \frac{|y|_1}{1 + L|y|_1}\pi_{s,x}(dy)ds \le \frac{1}{\rho t}(\|x\|_{-1}^2 + Ct), \quad \forall\, t \ge 0.$$

Letting $t \to \infty$ and recalling that μ is a weak limit of (μ_t) and that the support of μ is included in L^1, yields

$$\int_{H^{-1}} \frac{|y|_1}{1 + L|y|_1}\mu(dy) \le \frac{C}{\rho}.$$

Now the conclusion follows letting $L \to \infty$. $\square$

Remark 7.5.2 The existence of an invariant measure for $d > 1$ remains open.

7.6   The Full Support of Invariant Measures and Irreducibility of Transition Semigroups

We come back to Eq. (3.1) under Hypothesis 4 with additive noise, that is

$$\begin{cases} dX(t) = \Delta\beta(X(t))dt + \sqrt{Q}\,dW(t), \\[2mm] X(0) = x \in H^{-1}, \end{cases} \tag{7.59}$$

where Q is a linear positive operator in H^{-1} such that $\sigma := \sqrt{Q} \in \mathscr{L}_2(H^{-1}, L^2)$ $(\subset \mathscr{L}_2(H^{-1}))$. These assumptions are fulfilled taking

$$Q = A^{-\gamma}, \text{ where } \gamma > \frac{d+2}{2}. \tag{7.60}$$

Let P_t, $t \geq 0$, be the corresponding transition semigroup

$$P_t\varphi(x) := \mathbb{E}[\varphi(X(t,x))], \quad \forall\, t \geq 0, \ x \in H^{-1}, \ \varphi \in B_b(H^{-1}). \tag{7.61}$$

We recall (see e.g. [50]) that a Borel probability measure on H^{-1} is said to have *full support* if does not vanish on nonempty open subsets of H^{-1}. If μ is an invariant measure for P_t, $t \geq 0$, this property is implied by the *irreducibility* of P_t, $t \geq 0$. We recall that the semigroup P_t, $t \geq 0$, is called *irreducible* if

$$\mathbb{P}\left(\|X(T,x_0) - x_1\|_{-1} \geq r\right) < 1, \tag{7.62}$$

for all $T > 0$, $r > 0$, $x_0, x_1 \in H^{-1}$.

The main result of this section is the following.

Theorem 7.6.1 *Assume that Hypothesis 4 is fulfilled with $\sigma(x) = \sqrt{Q} = A^{-\gamma/2}$ and $\gamma = \frac{d+2}{2}$. Then the transition semigroup (7.61) is irreducible.*

As a consequence we have

Corollary 7.6.2 *The invariant measures given by Theorems 7.2.2, 7.3.1, 7.4.1 and 7.5.1 have full support in H^{-1}.*

To prove Theorem 7.6.1 we need some preliminaries. Set $B = \sqrt{Q}, E = H^{-1-\gamma}$ and consider for each $\epsilon > 0$ the deterministic porous media equation

$$\begin{cases} \dfrac{d}{dt}\, Y_\epsilon = \Delta\beta_\epsilon(Y_\epsilon) + Bu, & t \in [0,T] \\[2mm] \beta_\epsilon(Y_\epsilon) + \epsilon Y_\epsilon \in H_0^1 & \text{in } [0,T] \\[2mm] Y_\epsilon(0) = y_0 \in H^{-1}, \end{cases} \tag{7.63}$$

where as usually, $\beta_\epsilon = \frac{1}{\epsilon}\left(1 - (1 + \epsilon\beta)^{-1}\right)$. We know (see e.g. [6]) that for each $u \in L^2(0,T;E)$ (that is $Bu \in L^2(0,T;H^{-1})$) and $y_0 \in L^2$ Eq. (7.63) has a unique solution $Y_\epsilon = Y_\epsilon^u \in C([0,T];H^{-1})$ such that

$$\frac{d}{dt}\, Y_\epsilon \in L^2(0,T;H^{-1}), \quad Y_\epsilon \in L^1((0,T) \times \mathcal{O}), \quad \beta_\epsilon(Y_\epsilon) \in L^2(0,T;H_0^1).$$

We consider also the equation

$$\begin{cases} \dfrac{d}{dt}\, y = \Delta\beta(y) + Bu, & t \in [0,T] \\[2mm] \beta(y(t)) \in H_0^1 & \forall\, t \in [0,T] \\[2mm] y(0) = y_0 \in H^{-1}, \end{cases} \tag{7.64}$$

which has a unique solution $y = y''$ with

$$\frac{d}{dt}\, y \in L^2(\delta, T; H^{-1}), \quad y \in L^1((0,T) \times \mathcal{O}), \quad \beta(y) \in L^2(\delta, T; H_0^1),$$

for all $\delta \in (0, T)$. (See [6, p. 166].)

We denote by $F : D(F) \subset H^{-1} \to H^{-1}$ the maximal monotone operator

$$\begin{cases} F(y) = -\Delta\beta(y), \\[2mm] D(F) = \{y \in H^{-1} \cap L^1 : \beta(y) \in H_0^1\}. \end{cases}$$

We note that under our assumptions we have $D(F) = H^{-1}$. Denote by F^0 the minimal section of F (if F is multivalued).

The following approximating controllability result is the main ingredient of the proof of Theorem 7.6.1.

Lemma 7.6.3 *For all $\eta > 0$, $y_0 \in H^{-1}$, $y_1 \in D(F)$ there is $u_\epsilon \in L^2(0, T; E)$ such that $\|y_\epsilon(T) - y_1\|_{-1} \le \eta$. Moreover*

$$u_\epsilon \to u \quad \text{strongly in } L^2(0, T; E), \tag{7.65}$$

for $\epsilon \to 0$, where

$$\|y''(T) - y_1\|_{-1} \le \eta \tag{7.66}$$

We have also

$$|u_\epsilon|_{L^2(0,T;E)} \le C\eta^{-\frac{1}{2}} \left(\|y_0 - y_1\|_{-1} + \|F^0 y_1\|_{-1}\right). \tag{7.67}$$

Proof Let $y_1 \in D(F)$. Then for each $\epsilon > 0$ the solution to equation

$$\begin{cases} \dfrac{d}{dt}\, z_\epsilon(t) = \Delta\beta_\epsilon(z_\epsilon(t)) - \rho \ \text{sign}\,(z_\epsilon(t) - y_1) \ni 0, \quad t \in [0, T] \\[4mm] z_\epsilon(0) = y_0, \end{cases} \tag{7.68}$$

belongs to $C([0, T]; H^{-1}) \cap L^2(0, T; H_0^1)$ and satisfies $z_\epsilon(T) = y_1$ for

$$\rho = \frac{1}{T}\, \|y_0 - y_1\|_{-1} + \|F^0 y_1\|_{-1}.$$

Here "sign" is the multivalued operator in H^{-1}

$$\text{sign } z = \begin{cases} \frac{z}{\|z\|_{-1}} & \text{if } z \neq 0, \\ \{y \in H^{-1} : \|y\|_{-1} \leq 1\} & \text{if } z = 0. \end{cases}$$

Here is the argument. Since this operator is maximal monotone and everywhere defined, the operator

$$z \rightarrow -\Delta \beta_\epsilon(z) + \rho \text{ sign } (z - y_1),$$

is maximal monotone in H^{-1} and so, problem (7.68) is well defined. Indeed, by (7.68) in virtue of the monotonicity of the operator $-\Delta \beta_\epsilon$ we have

$$\frac{1}{2} \frac{d}{dt} \|z_\epsilon(t) - y_1\|_{-1}^2 + \rho \|z_\epsilon(t) - y_1\|_{-1}$$

$$\leq \|\beta_\epsilon(y_1)\|_1 \|z_\epsilon(t) - y_1\|_{-1}$$

$$\leq \|F^0 y_1\| \|z_\epsilon(t) - y_1\|_{-1}, \quad t > 0,$$

which implies $z_\epsilon(T) = y_1$ for $t \geq T$, as claimed.

Now we set $v_\epsilon := \rho \text{ sign } (z_\epsilon(t) - y_1)$. Since $z_\epsilon \rightarrow z$ for $\epsilon \rightarrow 0$ in $C([0, T]; H^{-1})$ where z is the solution to the problem

$$\begin{cases} \frac{d}{dt} z(t) = \Delta \beta(z(t)) - \rho \text{ sign } (z(t) - y_1) \ni 0, & \text{a.e } t \in [0, T] \\ z(0) = y_0, \end{cases}$$

and as easily seen $t \rightarrow \|z_\epsilon(t) - y_1\|_{-1}$ is monotonically decreasing, we infer that for $\epsilon \rightarrow 0$

$$v_\epsilon \rightarrow v \quad \text{strongly in } L^2(0, T; H^{-1}), \tag{7.69}$$

where $v(t) = \rho \text{ sign } (z(t) - y_1)$ for $t \in [0, T]$ and $z(T) = y_1$.

Let now $\eta > 0$ be arbitrary but fixed. We consider the minimization problem

$$\min \{|Bu - v_\epsilon|_{L^2(0,T;H^{-1})}^2 + \eta |u|_{L^2(0,T;E)}^2\}, \tag{7.70}$$

which clearly has a unique solution u_ϵ^η.

We have therefore

$$B^*(Bu_\epsilon^\eta - v_\epsilon) + \eta u_\epsilon^\eta = 0, \tag{7.71}$$

which yields by (7.71)

$$|Bu_\epsilon^\eta - v_\epsilon|_{L^2(0,T;H^{-1})}^2 + \eta|u_\epsilon^\eta|_{L^2(0,T;E)}^2 \le |v_\epsilon|_{L^2(0,T;H^{-1})}^2, \tag{7.72}$$

and

$$|B^*(Bu_\epsilon^\eta - v_\epsilon)|_{L^2(0,T;H^{-1})}^2 \le \sqrt{\eta}\,|v_\epsilon|_{L^2(0,T;H^{-1})}^2. \tag{7.73}$$

On the other hand, by (7.69) and (7.71) we see that

$$\lim_{\epsilon \to 0} u_\epsilon^\eta = u^\eta \quad \text{weakly in } L^2(0,T;E) \tag{7.74}$$

where $B^*(Bu^\eta - v) + \eta u^\eta = 0$. This yields

$$|Bu^\eta - v|_{L^2(0,T;H^{-1})}^2 + \eta|u^\eta|_{L^2(0,T;E)}^2 \le |v_\epsilon|_{L^2(0,T;H^{-1})}^2, \tag{7.75}$$

and by (7.73),

$$|B^*(Bu^\eta - v)|_{L^2(0,T;H^{-1})}^2 \le \sqrt{\eta}\,|v|_{L^2(0,T;H^{-1})}^2. \tag{7.76}$$

Replacing $\{u^\eta\}$ by a convex combination, we may assume that for $\eta \to 0$

$$Bu^\eta \to v \text{ strongly in } L^2(0,T;H^{-1}), \tag{7.77}$$

Indeed, $\{Bu^\eta - v\}$ is bounded and therefore weakly convergent on a subsequence to $\zeta \in L^2(0,T;H^{-1})$. Since by (7.76) $B^*\zeta = 0$ we infer that $\zeta = 0$ and so, (7.77) follows via Mazur's theorem. Now we have

$$|Bu_\epsilon^\eta - v_\epsilon|_{L^2(0,T;H^{-1})} \le |Bu_\epsilon^\eta - Bu^\eta|_{L^2(0,T;H^{-1})}$$

$$+|Bu^\eta - v|_{L^2(0,T;H^{-1})} + |v - v_\epsilon|_{L^2(0,T;H^{-1})}$$

$$\le \delta_1(\eta) + \delta_2(\epsilon) + |Bu_\epsilon^\eta - Bu^\eta|_{L^2(0,T;H^{-1})},$$

where $\delta_i(r) \to 0$ as $r \to 0$, $i = 1,2$. Let η be fixed and choose $\epsilon > 0$ such that $\delta_2(\epsilon) \le \delta_1(\eta)$.

Then we have

$$|Bu_\epsilon^\eta - v_\epsilon|_{L^2(0,T;H^{-1})} \le 3\delta_1(\eta), \quad \text{for } 0 < \epsilon < \epsilon_0(\eta)$$

and so $Y_\epsilon^\eta = y^{u_\epsilon^\eta}$ satisfies

$$\|Y_\epsilon^\eta(T) - z_\epsilon(T)\|_{-1} = \|Y_\epsilon^\eta(T) - y_1|_{-1} \le 3\delta_1(\eta)T.$$

Therefore redefining η by $\delta_1(\eta)$ we see that $u_\epsilon = u_\epsilon^\eta$ and u^η satisfies (7.65), (7.66). As regard to (7.67), it follows from estimates for v_ϵ and v. Indeed, by the first part of the proof we know that

$$\|v_\epsilon(t)\|_{-1} = \rho = \|F^0 y_1\|_{-1} + T^{-1}\|y_0 - y_1\|_{-1}, \quad \forall t \in [0, T]$$

and by (7.72)

$$|u_\epsilon^\eta|_{L^2(0,T;E)} \le \eta^{-\frac{1}{2}} \|v_\epsilon\|_{L^2(0,T;H^{-1})} \quad \forall t \in [0, T],$$

as claimed   □

We are now in position to prove Theorem 7.6.1.

Proof Let X_ϵ be the solution to the approximating equation ($\epsilon > 0$)

$$\begin{cases} dX_\epsilon(t) = \Delta\beta_\epsilon(X_\epsilon(t))dt + \sqrt{Q}\,dW(t), \\ X_\epsilon(0) = x \in H^{-1}, \end{cases} \tag{7.78}$$

Clearly it suffices to prove (7.62) for $x_0, x_1 \in D(F)$. Subtracting the latter equation from (7.63) where $u = u_\epsilon$ we obtain

$$\begin{cases} d(X_\epsilon - Y_\epsilon) = \Delta(\beta_\epsilon(X_\epsilon) - \beta_\epsilon(Y_\epsilon))dt + \sqrt{Q}\,(dW(t) - u_\epsilon dt), \\ (X_\epsilon - Y_\epsilon)(0) = 0. \end{cases}$$

This yields

$$X_\epsilon(t) - Y_\epsilon(t) + \theta_\epsilon(t) = \sqrt{Q}\,(W(t) - \tilde{v}_\epsilon d(t)), \tag{7.79}$$

where

$$\theta_\epsilon(t) = -\Delta \int_0^t (\beta_\epsilon(X_\epsilon(s)) - \beta_\epsilon(Y_\epsilon(s))ds$$

and

$$\tilde{v}_\epsilon(t) = \int_0^t u_\epsilon(s)ds.$$

This yields

$$\int_0^t \langle \beta_\epsilon(X_\epsilon(s)) - \beta_\epsilon(Y_\epsilon(s)), \theta_\epsilon(s)\rangle_2\, ds = \frac{1}{2}\,\|\theta_\epsilon(t)\|_{-1}^2$$

and so, by (7.79) we have

$$\int_0^t \langle \beta_\epsilon(X_\epsilon(s)) - \beta_\epsilon(Y_\epsilon(s)), \theta_\epsilon(s) \rangle_2 \, ds + \frac{1}{2} \, \|\theta_\epsilon(t)\|_{-1}^2$$

$$= \int_0^t \langle \beta_\epsilon(X_\epsilon(s)) - \beta_\epsilon(Y_\epsilon(s)), \sqrt{Q} \, W(s) - \tilde{v}_\epsilon \rangle_2 ds, \quad \forall \, t \in [0, T]$$

and therefore

$$\frac{1}{2} \, \|\theta_\epsilon(t)\|_{-1}^2$$

$$\leq \int_0^t \|\beta_\epsilon(X_\epsilon(s)) - \beta_\epsilon(Y_\epsilon(s))\|_1 \, ds \, \| \sqrt{Q} \, W - \tilde{v}_\epsilon \|_{C([0,T];\overline{\mathcal{O}})}.$$

(7.80)

Now by (7.63) where $y_0 = x$ and $u = u_\epsilon$ we see that (by multiplying with $Y_\epsilon - x$ in H^{-1})

$$\int_0^T \int_{\mathcal{O}} j_\epsilon(Y_\epsilon(s)) \, dt \, d\xi \leq C \left(\int_{\mathcal{O}} j(x) d\xi + |u_\epsilon|_{L^2(0,T;E)}^2 \right),$$

where $\nabla j_\epsilon = \beta_\epsilon$. Taking into account that

$$\beta_\epsilon(Y_\epsilon)(Y_\epsilon - \theta) \geq j_\epsilon(Y_\epsilon) - j_\epsilon(\theta), \quad \forall \, \theta \in \mathbb{R},$$

we get $\mathbb{P}$-a.s.

$$\int_0^T |\beta_\epsilon(Y_\epsilon)|_{L^1(\mathcal{O})} \leq C(|j(x)|_{L^1(\mathcal{O})} + |u_\epsilon|_{L^2(0,T;E)}^2 + 1) \leq C_1$$

(7.81)

Similarly we have by (7.78)

$$\int_0^t \int_{\mathcal{O}} |\beta_\epsilon(X_\epsilon)| ds \, d\xi \leq \int_0^t \int_{\mathcal{O}} \beta_\epsilon(X_\epsilon) X_\epsilon \, ds \, d\xi + C.$$

(7.82)

Now if we write (7.79) as

$$d(X_\epsilon - \sqrt{Q} \, W + \tilde{v}_\epsilon) - \Delta \beta_\epsilon(X_\epsilon) dt - u_\epsilon dt = 0$$

and multiply scalarly in H^{-1} by $(X_\epsilon - \sqrt{Q}\, W + \tilde{v}_\epsilon)$ we get

$$\int_0^t \int_{\mathcal{O}} \beta_\epsilon(X_\epsilon)(X_\epsilon - \sqrt{Q}\, W + \tilde{v}_\epsilon)\, ds\, d\xi$$

$$= \frac{1}{2}\, \|X_\epsilon(t) - \sqrt{Q}\, W(t) + \tilde{v}_\epsilon(t)\|_{-1}^2$$

$$+ \int_0^t \langle X_\epsilon(s) - \sqrt{Q}\, W(s) + \tilde{v}_\epsilon(s), u_\epsilon(s)\rangle_{-1}.$$

Hence by (7.82) we have

$$\int_0^t \int_{\mathcal{O}} |\beta_\epsilon(X_\epsilon)|\, ds\, d\xi + \frac{1}{2}\, \|(X_\epsilon(t) - \sqrt{Q}\, W(t) + \tilde{v}_\epsilon(t))\|_{-1}^2$$

$$\leq \int_0^t \int_{\mathcal{O}} \beta_\epsilon(X_\epsilon)(\sqrt{Q}\, W - \tilde{v}_\epsilon)\, ds\, d\xi + C. \tag{7.83}$$

Now for each η we have

$$\sup_{0 \leq s \leq t} \|\sqrt{Q}\, W(s) - \tilde{v}(s)\|_{C([0,T];\overline{\mathcal{O}})} \leq \eta^2 \quad \text{on } \Omega_\eta \subset \Omega,$$

where $\mathbb{P}(\Omega_\eta) > 0$. (This happens because the law of $\sqrt{Q}\, W$ is a nondegenerate Gaussian measure in $C([0,T]; \overline{\mathcal{O}})$.)

Since $\widetilde{v_\epsilon} \to \tilde{v}$ as $\epsilon \to 0$, it follows by (7.83) that

$$\int_0^t \int_{\mathcal{O}} |\beta_\epsilon(X_\epsilon)|\, ds\, d\xi \leq \eta^2 \int_0^t \int_{\mathcal{O}} |\beta_\epsilon(X_\epsilon)|\, ds\, d\xi + C(x_0, x_1), \quad \text{on } \Omega_\eta$$

and substituting along with (7.83) into (7.81) we get

$$\frac{1}{2}\, \|\theta_\epsilon(t)\|_{-1}^2 \leq C(x_0, x_1)(1 - \eta^2)^{-1}\eta^2, \quad \text{on } \Omega_\eta,$$

for ϵ sufficiently small. Hence by (7.80) we obtain that

$$\|X_\epsilon(T) - Y_\epsilon(T)\|_{-1} \leq \|\sqrt{Q}\, W(T) - \tilde{v}_\epsilon(T)\|_{-1/2}$$

$$+ C_1(x_0, x_1)(1 - \eta^2)^{-1}\eta, \quad \forall\, \epsilon > 0.$$

Letting $\epsilon \to 0$ we get on Ω_η,

$$\|X_\epsilon(T) - x_1\|_{-1} \leq \eta(1 + C_2(x_0, x_1)).$$

If we choose $\eta > 0$ such that $\eta + C_2(x_0, x_1)) < r$ we see that $\|X_\epsilon(T) - x_1\|_{-1} \leq r$ and consequently

$$\mathbb{P}(\|X_\epsilon(T) - x_1\|_{-1} > r) \leq 1 - \mathbb{P}(\Omega_\eta) < 1.$$

This completes the proof.   □

7.7   Comments and Bibliographical Remarks

For the general theory on transition semigroups and associated invariant measures we refer to the books [46, 50]. For the Stefan problem, Theorem 7.3.1 was proved in [10]. Existence and uniqueness for an invariant measure for slow diffusions was proved in [53]. For fast diffusions, existence of an invariant measure was proved in [11] and the uniqueness in [77].

In the case of slow diffusions porous media equation Theorem 7.6.1 was proved in [20]. It should be said, however, that the method of proof, based on controllability of the corresponding deterministic equation, extends to a more general class of nonlinear partial differential stochastic equations.

References

1. R. Adams, *Sobolev Spaces* (Academic Press, San Francisco, 1975)
2. S.A. Agmon, *Lectures on Elliptic Boundary Value Problems* (Van Nostrand, Princeton, 1965)
3. S.N. Antontsev, J.F. Diaz, S. Shmarev, *Energy Methods for Free Boundary Problems* (Birkhäuser, Basel, 2002)
4. P. Bak, K. Chen, Self-organized criticality. Sci. Am. **264**, 40 (1991)
5. P. Bak, C. Tang, K. Wiesenfeld, Phys. Rev. Lett. **59**, 381–394 (1987); Phys. Rev. A **38**, 364–375 (1988)
6. V. Barbu, *Nonlinear Differential Equations of Monotone Type in Banach Spaces* (Springer, New York, 2010)
7. V. Barbu, Self-organized criticality and convergence to equilibrium of solutions to nonlinear diffusion problems. Annu. Rev. Control. JARAP **340**, 52–61 (2010)
8. V. Barbu, A variational approach to stochastic nonlinear parabolic problems. J. Math. Anal. Appl. **384**, 2–15 (2011)
9. V. Barbu, Optimal control approach to nonlinear diffusion equations driven by Wiener noise. J. Optim. Theory Appl. **153**, 1–26 (2012)
10. V. Barbu, G. Da Prato, The two phase stochastic Stefan problem. Probab. Theory Relat. Fields **124**, 544–560 (2002)
11. V. Barbu, G. Da Prato, Invariant measures and the Kolmogorov equation for the stochastic fast diffusion equation. Stoch. Process. Appl. **120**(7), 1247–1266 (2010)
12. V. Barbu, G. Da Prato, Ergodicity for the phase-field equations perturbed by Gaussian noise. Infinite Dimen. Anal. Quantum Probab. Relat. Top. **14**(1), 35–55 (2011)
13. V. Barbu, C. Marinelli, Strong solutions for stochastic porous media equations with jumps. Infinite Anal. Quantum Probab. Relat. Fields **12**, 413–426 (2009)
14. V. Barbu, Th. Precupanu, Convexity and Optimization in Banach Spaces (Springer, New York, 2011)
15. V. Barbu, M. Röckner, On a random scaled porous media equations. J. Differ. Equ. **251**, 2494–2514 (2011)
16. V. Barbu, M. Röckner, Localization of solutions to stochastic porous media equations: finite speed of propagation. Electron. J. Probab. **17**, 1–11 (2012)
17. V. Barbu, M. Röckner, Stochastic porous media and self-organized criticality: convergence to the critical state in all dimensions. Commun. Math. Phys. **311**(2), 539–555 (2012)
18. V. Barbu, M. Röckner, Stochastic variational inequalities and applications to the total variational flow perturbed by linear multiplicative noise. Arch. Ration. Mech. Anal. **209**, 797–834 (2013)

19. V. Barbu, M. Röckner, An operatorial approach to stochastic partial differential equations driven by linear multiplicative noise. J. Eur. Math, Soc. **17**, 1789–1815 (2015)

20. V. Barbu, V. Bogachev, G. Da Prato, M. Röckner, Weak solutions to the stochastic porous media equation via Kolmogorov equations: the degenerate case. J. Funct. Anal. **237**, 54–75 (2006)

21. V. Barbu, G. Da Prato, M. Röckner, Existence and uniqueness of nonnegative solutions to the stochastic porous media equation. Indiana Univ. Math. J. **57**(1), 187–212 (2008)

22. V. Barbu, G. Da Prato, M. Röckner, Existence of strong solutions for stochastic porous media equation under general monotonicity conditions. Ann. Probab. **37**(2), 428–452 (2009)

23. V. Barbu, G. Da Prato, M. Röckner, Finite time extinction for solutions to fast diffusion stochastic porous media equations. C. R. Acad. Sci. Paris, Ser. I **347**, 81–84 (2009)

24. V. Barbu, G. Da Prato, M. Röckner, Stochastic porous media equation and self-organized criticality. Commun. Math. Phys. **285**, 901–923 (2009)

25. V. Barbu, G. Da Prato, M. Röckner, Finite time extinction of solutions to fast diffusion equations driven by linear multiplicative noise. J. Math. Anal. Appl. **389**, 147–164 (2012)

26. V. Barbu, M. Röckner, F. Russo, The stochastic porous media equations in $\mathbb{R}^d$. J. Math. Pures Appl. (9), **103**(4), 1024–1052 (2015)

27. V. Barbu, M. Röckner, F. Russo, A stochastic Fokker-Planck equation and double probabilistic representation for the stochastic porous media type equation. arXiv:1404.5120

28. J.G. Berryman, C.J. Holland, Nonlinear diffusion problems arising in plasma physics. Phys. Rev. Lett. **40**, 1720–1722 (1978)

29. J.G. Berryman, C.J. Holland, Asymptotic behavior of the nonlinear diffusion equations. J. Math. Phys. **54**, 425–426 (1983)

30. Ph. Blanchard, M. Röckner, F. Russo, Probabilistic representation for solutions of an irregular porous media type equation. Ann. Probab. **38**(5), 1870–1900 (2010)

31. V. Bogachev, G. Da Prato, M. Röckner, Invariant measures of stochastic porous medium type equations. Dokl. Math. Russ. Acad. Sci. **396**(1), 7–11 (Russian) (2004)

32. V. Bogachev, G. Da Prato, M. Röckner, Fokker–Planck equations and maximal dissipativity for Kolmogorov operators with time dependent singular drifts in Hilbert spaces. J. Funct. Analysis **256**, 1269–1298 (2009)

33. V. Bogachev, G. Da Prato, M. Röckner, Existence and uniqueness of solutions for Fokker–Planck equations on Hilbert spaces. J. Evol. Equ. **10**(3), 487–509 (2010)

34. V. Bogachev, G. Da Prato, M. Röckner, Uniqueness for solutions of Fokker–Planck equations on infinite dimensional spaces. Commun. Partial Differ. Equ. **36**, 925–939 (2011)

35. H. Brezis, *Operatéurs Maximaux Monotones* (North-Holland, Amsterdam, 1973)

36. H. Brezis, *Functional Analysis, Sobolev Spaces and Partial Differential Equations* (Springer, New York, 2010)

37. H. Brezis, M. Crandall, Uniqueness of solutions of the initial-value problem for $u_t - \Delta\varphi(u) = 0$. J. Math. Pures Appl. **58**, 153–163 (1979)

38. Z. Brzezniak, J.van Neerven, M. Veraar, L. Weis, L. Itô's formula in UMD Banach spaces and regularity of solutions of the Zakai equation. J. Differ. Equ. **245**(1), 30–58 (2008)

39. R. Cafiero, V. Loreto, L. Pietronero, A.Vespigniani, S. Zapperi, Local rigidity and self-organized criticality for Avalanches. Europhys. Lett. **29**(2), 111—116 (1995)

40. P. Cannarsa, G. Da Prato, Invariance for stochastic reaction-diffusion equations. Evol. Equ. Control Theory **1**(1), 43–56 (2012)

41. J.M. Carlson, G.H. Swindle, Self-organized criticality: sand piles, singularities and scaling. Proc. Natl. Acad. Sci. USA **92**, 6710–6719 (1995)

42. J.M. Carlson, E.R. Changes, E.R. Grannan, G.H. Swindle, Self-organized criticality and singular diffusions. Phys. Rev. Lett. **65**, 2547–2550 (1990)

43. S. Cerrai, A Hille–Yosida theorem for weakly continuous semigroups. Semigroup Forum **49**, 349–367 (1994)

44. I. Ciotir, A Trotter type theorem for nonlinear stochastic equations in variational formulation and homogenization. Differ. Integr. Equ. **24**, 371–388 (2011)

45. I. Ciotir, Convergence of solutions for stochastic porous media equations and homegenization. J. Evol. Equ. **11**, 339–370 (2011)
46. G. Da Prato, *Kolmogorov Equations for Stochastic PDEs* (Birkhäuser, Basel, 2004)
47. G. Da Prato, M. Röckner, Weak solutions to stochastic porous media equations. J. Evol. Equ. **4**, 249–271 (2004)
48. G. Da Prato, M. Röckner, Invariant measures for a stochastic porous medium equation, in *Stochastic analysis and related topics in Kyoto*. Advanced Studies in Pure Mathematics, vol. 41 (The Mathematical Society of Japan, Tokyo, 2004), pp. 13–29
49. G. Da Prato, M. Röckner, Well posedness of Fokker–Planck equations for generators of time-inhomogeneous Markovian transition probabilities. Rend. Lincei Mat. Appl. **23**(4), 361–376 (2012)
50. G. Da Prato, J. Zabczyk, *Ergodicity for Infinite Dimensional Systems*. London Mathematical Society Lecture Notes, vol. 229 (Cambridge University Press, Cambridge, 1996)
51. G. Da Prato, J. Zabczyk, *Stochastic Equations in Infinite Dimensions*. Encyclopedia of Mathematics and Its Applications, 2nd edn. (Cambridge University Press, Cambridge, 2014)
52. G. Da Prato, A. Debussche, B. Goldys, Invariant measures of non symmetric dissipative stochastic systems. Probab. Theory Relat. Fields **123**(3), 355–380 (2002)
53. G. Da Prato, M. Röckner, B.L. Rozovskii, F. Wang, Strong solutions of stochastic generalized porous media equations: existence, uniqueness, and ergodicity. Commun. Partial Differ. Equ. **31**(1–3), 277–291 (2006)
54. J.I. Diaz, L. Veron, Local vanishing properties of solutions of elliptic and parabolic quasilinear equations. Trans. Am. Math. Soc. **290**, 787–814 (1985)
55. C.M. Elliot, *Free Boundary Problems* (Pitman, London, 1982)
56. C.M. Elliot, J.R. Okendon, *Weak and Variational Methods for Moving Boundary Problems*. Pitman Research Notes in Mathematics, vol. 59 (Pitman, Boston, 1982)
57. L. Evans, R. Gariepy, *Measure Theory and Fine Properties of Functions*. Studies in Advanced Mathematics (CRC Press, Boca Raton, 1992)
58. M. Fukushima, Y. Oshima, M. Takeda, *Dirichlet Forms and Symmetric Markov Processes*. de Gruyter Studies in Mathematics, vol. 19, Second revised and extended edn. (Walter de Gruyter & Co., Berlin, 2011)
59. B. Gess, Finite speed of propagation for stochastic porous media equations. SIAM J. Math. Anal. **45**(5), 2734–2766 (2013)
60. B. Gess, Random attractors for stochastic porous media equations perturbed by space linear multiplicative noise. Ann. Probab. **42**(2), 818–864 (2014)
61. B. Gess, Finite time extinction for signfast diffusions and self-organized criticality. Commun. Math. Phys. **335**, 309–344 (2015)
62. B. Gess, M. Röckner, Singular-degenerate multivalued stochastic fast diffusion equations. SIAM J. Math. Anal. **47**(5), 4058–4090 (2015)
63. B. Gess, M. Röckner, Stochastic variational inequalities and regularity for degenerate stochastic partial differential equations. Trans. Am. Math. Soc. (to appear). arXiv:1405.5866
64. B. Gess, J. Tölle, Multivalued stochastic evolution inclusions. J. Math. Pures Appl. **101**(6), 789–827 (2013)
65. B. Gess, J. Tölle, Stability of solutions to stochastic partial differential equations, 1–39 (2015). arXiv:1506.01230
66. D. Grieser, Uniform bounds for eigenfunctions of the Laplacian on manifolds with boundary. Commun. Partial Differ. Equ **27**(7–8), 1283–1299 (2002)
67. L. Hörmander, *The Analysis of Linear Partial Differential Operators* (Springer, New York, 1963)
68. J.U. Kim, On the stochastic porous medium equation. J. Differ. Equ. **220**, 163–194 (2006)
69. K. Kim, C. Mueller, R. Sowers, A stochastic moving boundary value problem. Ill. J. Math. **54**(3), 927–962 (2010)
70. K. Kim, Z. Zheng, R. Sowers, A stochastic Stefan problem. J. Theor. Probab. **25**(4), 1040–1080 (2012)

71. N. Krylov, Itô's formula for the L^p-norm of stochastic W_p^1-valued processes. Probab. Theory. Relat. Fields **147**, 583–605 (2010)
72. N. Krylov, B.L. Rozovskii, *Stochastic evolution equations*. J. Sov. Math. **16**, 1233–1277 (1981)
73. P. Li, S.T. You, On the Schrödinger equation and the eigenvalue problem. Commun. Math. Phys. **88**, 309–318 (1983)
74. J.L. Lions, *Quelques Méthodes de Solutions de Problèmes Non Linéaires* (Dunod/Gauthier-Villars, Paris, 1969)
75. R.S. Liptser, A.N. Shiryayev, *Theory of Martingales*. (Translated from the Russian by K. Dzjaparidze [Kacha Dzhaparidze]). Mathematics and Its Applications (Soviet Series), vol. 49 (Kluwer, Dordrecht, 1989)
76. W. Liu, M. Röckner, *Stochastic Partial Differential Equations: An Introduction*. Universitext. (Springer, Cham, 2015)
77. W. Liu, J. Tölle, Existence and uniqueness of invariant measures for stochastic evolution equations with weakly dissipative drifts. Electronic Commun. Probab. **16**, 447–457 (2011)
78. S. Lototsky, A random change of variables and applications to the stochastic porous medium equation with multiplicative time noise. Commun. Stoch. Anal. **1**(3), 343–355 (2007)
79. G. Marinoschi, *Functional Approach to Nonlinear Models of Water Flows in Soils* (Springer, Dordrecht, 2006)
80. E. Pardoux, *Équations aux dérivées partielle stochastiques nonlineaires monotones*, Thèse, Paris, 1972
81. I.J. Pedron, R.S. Mendes, T.J. Buratto, L.C. Malacarne, E.K. Lenzi, Logarithmic diffusion and porous media equations: a unified description. Phys. Rev. E **72**, 031106 (2005)
82. C. Prevot, M. Röckner, *A Concise Course on Stochastic Partial Differential Equations*. Lecture Notes in Mathematics, vol. 1905 (Springer, Berlin, 2007)
83. E. Priola, *On a class of Markov type semigroups in spaces of uniformly continuous and bounded functions*. Stud. Math. **136**, 271–295 (1999)
84. J. Ren, M. Röckner, F.Y. Wang, Stochastic generalized porous media and fast diffusion equations. J. Differ. Equ. **238**(1), 118–152 (2007)
85. D. Revuz, M. Yor, *Continuous Martingales and Brownian Motion*, 3rd edn. (Springer, Berlin, 1999)
86. M. Röckner, F.Y. Wang, Non-monotone stochastic generalized porous media equations. J. Differ. Equ. **245**(12), 3898–3935 (2008)
87. M. Röckner, F.Y. Wang, General extinction results for stochastic partial differential equations and applications. J. Lond. Math. Soc. **87**(2), 3943–3962 (2013)
88. Ph. Rosneau, Fast and super fast diffusion processes. Phys. Rev. Lett. **74**, 1057–1059 (1995)
89. J.L.Vazquez, *Smoothing and Decay Estimates for Nonlinear Diffusion Equations. Equations of Porous Medium Type*. Oxford Lecture Series in Mathematics and Its Applications, vol. 33 (Oxford University Press, Oxford, 2006)
90. J.L. Vazquez, J.R. Esteban, A. Rodriguez, The fast diffusion equation with logarithmic nonlinearity and the evolution of conformal metrics in the plane. Adv. Differ. Equ. **1**, 21–50 (1996)
91. H. Weyl, Das asymptotische Verteilungsgesetz der Eigenwerte linearer partieller Differential-gleichungen (mit einer Anwendung auf die Theorie der Hohlraumstrahlung). (German) Math. Ann. **71**(4), 441–479 (1912)

Index

H-valued adapted processes, 9

Birkhoff–Von Neumann theorem, 174
Brezis–Ekeland principle, 93
Burkholder–Davis–Gundy, 9

conjugate, 15
convex and lower semicontinuous, 14
convex integrals, 117
cylindrical Wiener process, 8

distributional solution, 50
Dunford–Pettis, 18

equi-integrable, 17
extinction probability, 65

fast diffusion, 2

Hilbert–Schmidt operators, 8

infinitesimal generator, 169
irreducibility, 188
Itô's formula for the L^p norm, 10
Itô's process, 10

Legendre transform, 15
Lipschitz continuous, 25
local martingale, 12
logarithmic diffusion equation, 88
low diffusion, 2
Luxemburg norm, 100

Markov transition semigroup, 168
martingale, 9
monotone, 12

noise, 2

Orlicz spaces, 99

porous media equation, v, 2
probability kernel, 169

rescaling approach, 63
Richard's equation, 111

sand-pile model, 3
self-organized criticality, 3
slow diffusions, vi
Sobolev–Gagliardo–Nirenberg theorem, 98
stochastic porous media equation, v
stochastic processes, 9

© Springer International Publishing Switzerland 2016
V. Barbu et al., *Stochastic Porous Media Equations*, Lecture Notes
in Mathematics 2163, DOI 10.1007/978-3-319-41069-2

subdifferential, 14
superfast diffusion, 2

temperate distributions, 134
transition semigroups, 167
two phase transition Stefan problem, 5

variational approach, 95

Yosida approximations, 13
Young function, 99

LECTURE NOTES IN MATHEMATICS  Springer

Editors in Chief: J.-M. Morel, B. Teissier;

Editorial Policy

1. Lecture Notes aim to report new developments in all areas of mathematics and their applications – quickly, informally and at a high level. Mathematical texts analysing new developments in modelling and numerical simulation are welcome.

   Manuscripts should be reasonably self-contained and rounded off. Thus they may, and often will, present not only results of the author but also related work by other people. They may be based on specialised lecture courses. Furthermore, the manuscripts should provide sufficient motivation, examples and applications. This clearly distinguishes Lecture Notes from journal articles or technical reports which normally are very concise. Articles intended for a journal but too long to be accepted by most journals, usually do not have this "lecture notes" character. For similar reasons it is unusual for doctoral theses to be accepted for the Lecture Notes series, though habilitation theses may be appropriate.

2. Besides monographs, multi-author manuscripts resulting from SUMMER SCHOOLS or similar INTENSIVE COURSES are welcome, provided their objective was held to present an active mathematical topic to an audience at the beginning or intermediate graduate level (a list of participants should be provided).

   The resulting manuscript should not be just a collection of course notes, but should require advance planning and coordination among the main lecturers. The subject matter should dictate the structure of the book. This structure should be motivated and explained in a scientific introduction, and the notation, references, index and formulation of results should be, if possible, unified by the editors. Each contribution should have an abstract and an introduction referring to the other contributions. In other words, more preparatory work must go into a multi-authored volume than simply assembling a disparate collection of papers, communicated at the event.

3. Manuscripts should be submitted either online at www.editorialmanager.com/lnm to Springer's mathematics editorial in Heidelberg, or electronically to one of the series editors. Authors should be aware that incomplete or insufficiently close-to-final manuscripts almost always result in longer refereeing times and nevertheless unclear referees' recommendations, making further refereeing of a final draft necessary. The strict minimum amount of material that will be considered should include a detailed outline describing the planned contents of each chapter, a bibliography and several sample chapters. Parallel submission of a manuscript to another publisher while under consideration for LNM is not acceptable and can lead to rejection.

4. In general, **monographs** will be sent out to at least 2 external referees for evaluation.

   A final decision to publish can be made only on the basis of the complete manuscript, however a refereeing process leading to a preliminary decision can be based on a pre-final or incomplete manuscript.

   Volume Editors of **multi-author works** are expected to arrange for the refereeing, to the usual scientific standards, of the individual contributions. If the resulting reports can be forwarded to the LNM Editorial Board, this is very helpful. If no reports are forwarded or if other questions remain unclear in respect of homogeneity etc, the series editors may wish to consult external referees for an overall evaluation of the volume.

5. Manuscripts should in general be submitted in English. Final manuscripts should contain at least 100 pages of mathematical text and should always include

   - a table of contents;
   - an informative introduction, with adequate motivation and perhaps some historical remarks: it should be accessible to a reader not intimately familiar with the topic treated;
   - a subject index: as a rule this is genuinely helpful for the reader.
   - For evaluation purposes, manuscripts should be submitted as pdf files.

6. Careful preparation of the manuscripts will help keep production time short besides ensuring satisfactory appearance of the finished book in print and online. After acceptance of the manuscript authors will be asked to prepare the final LaTeX source files (see LaTeX templates online: https://www.springer.com/gb/authors-editors/book-authors-editors/manuscriptpreparation/5636) plus the corresponding pdf- or zipped ps-file. The LaTeX source files are essential for producing the full-text online version of the book, see http://link.springer.com/bookseries/304 for the existing online volumes of LNM). The technical production of a Lecture Notes volume takes approximately 12 weeks. Additional instructions, if necessary, are available on request from lnm@springer.com.

7. Authors receive a total of 30 free copies of their volume and free access to their book on SpringerLink, but no royalties. They are entitled to a discount of 33.3 % on the price of Springer books purchased for their personal use, if ordering directly from Springer.

8. Commitment to publish is made by a *Publishing Agreement*; contributing authors of multiauthor books are requested to sign a *Consent to Publish form*. Springer-Verlag registers the copyright for each volume. Authors are free to reuse material contained in their LNM volumes in later publications: a brief written (or e-mail) request for formal permission is sufficient.

Addresses:
Professor Jean-Michel Morel, CMLA, École Normale Supérieure de Cachan, France
E-mail: moreljeanmichel@gmail.com

Professor Bernard Teissier, Equipe Géométrie et Dynamique,
Institut de Mathématiques de Jussieu – Paris Rive Gauche, Paris, France
E-mail: bernard.teissier@imj-prg.fr

Springer: Ute McCrory, Mathematics, Heidelberg, Germany,
E-mail: lnm@springer.com

Printed in the United States
By Bookmasters

Printed in the United States
By Bookmasters